Theoretical and Mathematical Physics

The series founded in 1975 and formerly (until 2005) entitled *Texts and Monographs in Physics* (TMP) publishes high-level monographs in theoretical and mathematical physics. The change of title to *Theoretical and Mathematical Physics* (TMP) signals that the series is a suitable publication platform for both the mathematical and the theoretical physicist. The wider scope of the series is reflected by the composition of the editorial board, comprising both physicists and mathematicians.

The books, written in a didactic style and containing a certain amount of elementary background material, bridge the gap between advanced textbooks and research monographs. They can thus serve as basis for advanced studies, not only for lectures and seminars at graduate level, but also for scientists entering a field of research.

T0205999

Martin Schlichenmaier

An Introduction
to Riemann Surfaces,
Algebraic Curves
and Moduli Spaces

Second Edition

With 21 Figures

 Springer

Prof. Dr. Martin Schlichenmaier
University of Luxembourg
Institute of Mathematics
162A, avenue de la Faiencerie
1511 Luxembourg City
Grand-Duchy of Luxembourg

Martin Schlichenmaier, *An Introduction to Riemann Surfaces, Algebraic Curves and Moduli Spaces*, Theoretical and Mathematical Physics (Springer, Berlin Heidelberg 2007) DOI 10.1007/b11501497

ISSN 0172-5998
ISBN 978-3-642-09027-1
e-ISBN 978-3-540-71175-9

Springer is a part of Springer Science+Business Media
springer.com
© Springer-Verlag Berlin Heidelberg 2007
Softcover reprint of the hardcover 2nd edition 2007

Cover design: eStudio Calamar, Girona/Spain

Preface to the Second Edition

During the second advent of string theory the first edition of this Springer Lecture Notes volume appeared in 1990. There was an increasing demand for physicists to learn more about the modern aspects of geometry. In particular, for a further development of the physical theory the notions mentioned in the title turned out to be of fundamental importance.

The first edition was based on lecture courses I gave at the Institute of Theoretical Physics, University of Karlsruhe, Germany. During this time the institute was headed by Prof. Julius Wess. Indeed, it was him who convinced me to write them up. Instead of repeating everything let me refer to the *Introduction to the 1st Edition* which can be found essentially without changes in this volume. For the motivation coming from the physics of these years see especially the introduction given by Prof. Ian McArthur.

Clearly, in the meantime Theoretical Physics developed further. Nevertheless, the geometric concepts introduced in the first edition (and even more) remain to be the very basic knowledge expected to be known by everybody working in related fields of theoretical physics. Consequently, the book is still in demand. But as it is out of print, the publisher suggested that I should prepare a revised and enlarged version of it. This offer I gratefully accepted.

As far as the revised part is concerned the changes are mainly due to removing typographical errors, making some unclear statements more precise, smoothening the presentation, etc. It was my clear aim to keep the structure and the partly informal style of the first edition.

In the meantime even more advanced geometric techniques are needed by physicists. From the many possible choices I took the following two additional topics: (1) the modern language and modern view of Algebraic Geometry and (2) Mirror Symmetry.

In Chaps. 11 and 12 I will describe the modern point of view of Algebraic Geometry. One of the basic ideas there is to replace the classical geometric space by the algebra of functions on the space or even more generally by the set of maps from this space to other spaces and vice versa, respectively. The geometry corresponds to some algebraic structure on the set of maps. These concepts were established by Grothendieck in the 1960s. What does one gain by this generalization? One of the advantages is that it is possible to extend the techniques of geometry to more general objects which do not have an obvious structure of a "classical space". Moreover, from the problem of the existence of moduli spaces classifying classical geometric objects one is led immediately outside of the category of "classical manifolds". This point of view was very fruitful in mathematics. See, e.g. the proof of the Weil conjectures by Pierre Deligne, Faltings' proof of Mordell's conjecture, Faltings' proof of the Verlinde formula, Wiles' proof of Fermat's Last Theorem, ... But it is also of importance in Theoretical Physics. There noncommutative space, quantum groups, and more involved objects show up. This is one of the reasons for the increasing interest in modern algebraic geometry among theoretical physicists. Furthermore, spaces with singularities are incorporated from the very beginning of the theory. In Chap. 11 local aspects are considered. The spectrum of a ring is introduced. It is shown how the concept of a point can be replaced by the concept of a homomorphism. Also the noncommutative case, of relevance for the quantum plane and for quantum groups, will be considered. In Sect. 12 global aspects, i.e. affine schemes and general schemes, will be introduced.

Mirror symmetry is another example for the very fruitful ongoing interaction between Theoretical Physics, Mathematical Physics and Fundamental Mathematics. This interaction works in both directions. On one hand, mathematics supplies the necessary tools and results for physicists. On the other hand, based on the physical theory quite unexpected mathematical relations are conjectured which turn out to be real challenges for mathematicians to be understood mathematically.

In very rough terms, the origin of mirror symmetry in physics might be described as follows. There are five models for superstring theories. Their common feature is that the dimension of spacetime equals ten. To obtain the four-dimensional spacetime, which we experience, the remaining six dimensions are thought to be very small and compact. This splitting is also called *string compactification.* The most appealing candidates for the compactified factors are some special compact complex three-dimensional manifold, the so-called Calabi-Yau threefolds. The physical models obtained by different compactifications show certain dualities. One of these (conjectured) dualities (relating type IIA string theory on one Calabi-Yau threefold to type IIB string theory on its conjectured mirror) is the mirror symmetry.

In Chaps. 13 and 14 the mathematical background of mirror symmetry and the mirror conjectures will be explained. To formulate mirror symmetry the theory of Kähler manifolds, Hodge decomposition and Calabi-Yau manifolds will be developed. Some general remarks on the physical background can be found in Sect. 13.1.

Indeed from the mathematical point of view there are different levels of mirror symmetry, respectively different mirror conjectures. The first version is the *Geometric Mirror Symmetry.* It is the conjecture that for certain classes of Calabi-Yau manifolds X there exist mirrors Y such that their Hodge diamonds, constituted by their individual Hodge numbers $h^{p,q}$, are mirror symmetric. More precisely, we get $h^{p,q}(X) = h^{n-p,q}(Y)$. The *Numerical Mirror Symmetry* gives a relation between the generating function for the number of rational curves of degree d on X and a function defined in terms of period integrals on the mirror Y. The third variant is *Kontsevich's Homological Mirror Conjecture*. It says that for mirror pairs (X, Y) of Calabi-Yau manifolds the derived Fukaya A_∞-category of Lagrangian submanifolds of X is isomorphic to the derived category of coherent sheaves of Y.

In Chap. 14 the geometric mirror conjecture is explained in detail. The numerical mirror conjecture is sketched for the quintic hypersurface in four-dimensional projective space. This is the classical example discovered by the physicists.

As, compared to the first edition, the total number of pages increased the following *Leitfaden* might be useful for the reader.

The first part of the book (Chaps. 1–7) should be considered as the very basic knowledge. It gives a concise introduction to the theory of Riemann surfaces, curves and their moduli spaces. It was my goal to keep the presentation as elementary as possible. After some fundamental definitions, for instance, the definitions of manifolds, differentials and other basic objects, we restrict ourselves to Riemann surfaces. First, we study their topology, beginning with their fundamental groups. As a fundamental example of homology theories we discuss simplicial homology. A Riemann surface is a manifold which locally looks like the complex numbers $\mathbb{C}$. Hence we have available the theory of analytic functions, differentials, and so on. If we restrict ourselves to compact Riemann surfaces, we get some very useful results such as the theorem of Riemann–Roch. Roughly speaking, this gives us informations concerning the existence of functions with prescribed behaviour at certain points. The next goal is to study the set of different analytic isomorphism classes of Riemann surfaces. Equipped with a geometric structure representing how the Riemann surfaces "appear in nature" this set is called a moduli space. One important step in doing this is to assign to every Riemann surface its Jacobian. This is a higher dimensional torus with the crucial property that it is embeddable into some projective space. Further, to study these moduli spaces we use the language of algebraic geometry, which we must first develop. In the first part of the book we use only the more classical concept of varieties.

The second part consists of Chaps. 8 and 9. It contains the more advanced basics and introduces more advanced tools like line bundles, the relations between line bundles and divisors, vector bundles, cohomology, the theorem of Riemann–Roch for line bundles, etc.

The following parts are divided in blocks which are essentially independent of each others. The reader can decide which parts he wants to study.

In Chap. 10 the Grothendieck–Riemann–Roch–Hirzebruch theorem is formulated. It is shown how to deduce the Riemann–Roch theorem from it. Furthermore, the Mumford isomorphism, of fundamental relevance for the geometry of the moduli space of curves and for the string partition function, is deduced. As prerequisites clearly the above two basic parts are needed.

The next block, consisting of Chaps. 11 and 12, deals with modern algebraic geometry. Its content was already roughly described above. To allow that these two chapters can be studied completely independent of the other chapters I repeated in these chapters all definitions needed. Nevertheless, as the objects there are rather abstract, an knowledge about the geometric objects presented in Part 1 will be of advantage.

Chapters 13 and 14 on mirror symmetry depend again on the two basic parts. A knowledge of Chap. 10 might be helpful.

The final appendix on p-adic numbers is completely independent from the rest of the book.

This book addresses not only theoretical physicists, mathematical physicists but also mathematicians. It should help the reader to enter the field of modern geometry. It was my goal to keep the style of the first edition. In several respects there will be mixture between different techniques and viewpoints. In this way beautiful relations between different approaches can be detected.

The book should be self-contained in the sense of the developed mathematical notions and understanding. For the proof of quite a number of central theorems (like Riemann–Roch, etc.) I mainly refer to the mathematical literature. For other mathematical facts I supply proofs. The reason for these proofs is either the fact that in these cases without the proof the mathematical statement cannot be fully understood or that they show how to work with the developed mathematical theory and to draw consequences from it.

The first edition grew out from several lecture courses. This influenced clearly the style of the presentation. Several times I first made a definition for the particular important special case under consideration. The intention was not to overload the reader with only definitions in the beginning. The most general definition was given later on in those parts of the book where it is really needed. At least to a certain extent it was my goal to keep the lecture style also for the added chapters of the second edition. In fact the part on modern algebraic geometry developed out from a write-up of lecture series which I gave at an autumn school of the Graduiertenkolleg *Mathematik im Bereich Ihrer Wechselwirkungen mit der Physik* of the Ludwig-Maximilians-Universität, München. It is a pleasure for me to thank Prof. Martin Schottenloher and Prof. Julius Wess for the invitation to present the lectures there.

Luxembourg,   *Martin Schlichenmaier*
September 2006

Preface to the First Edition

This is a write-up of a lecture course I gave in the fall term 1987/1988 at the Institute of Theoretical Physics of the University of Karlsruhe. Ideas from algebraic and analytic geometry are becoming more and more useful in Theoretical Physics, particularly in relation to string theories and related subjects, and so I was asked by Prof. Julius Wess to give an introductory lecture course for physicists on these exciting mathematical ideas.

Of course, there are many different strategies available for putting together such a course. And there are as many arguments for as there are against the choice of subjects made here and the way to proceed. Certainly my personal preference played a part in this choice, but, nevertheless, I hope the audience of the lectures and now the readers of these notes, if interested, will be able to study the appropriate mathematics books after the lecture in a particular subject.

Let me give a brief review of the lectures. After some fundamental definitions, for instance, of manifolds, differentials and other basic objects, we restrict ourselves to Riemann surfaces. First, we study their topology, beginning with their fundamental groups. As a nice example of homology theories we discuss simplicial homology. A Riemann surface is a manifold which locally looks like the complex numbers $\mathbb{C}$. Hence we have available the theory of analytic functions, differentials and so on. If we restrict ourselves to compact Riemann surfaces, we get some very useful results such as the theorem of Riemann–Roch. Roughly speaking this gives us information concerning the existence of functions with prescribed behaviour at certain points.

In Polyakov bosonic string theory the partition function is given by "integrating" over all topologies and metrics on an orientable compact surface. By gauge invariance the integration is eventually reduced to the set of different analytic isomorphism classes of Riemann surfaces. Equipped with a certain geometric structure this set is called a moduli space. As you see, it is not enough to study one fixed (although arbitrary) Riemann surface, we have to study all of them together. One important step in doing this is to assign to every Riemann surface its Jacobian. This is a higher dimensional torus with the crucial property that it is embeddable into some projective space. Further, to study these moduli spaces we use the language of algebraic geometry, which we must first develop. Here we use only the more classical concept of varieties. With algebraic geometry we have a very powerful machinery at our disposal for studying the geometry of the moduli space.

We develop the theory of vector bundles, sheaves and sheaf cohomology and deduce the Mumford isomorphism on the moduli space starting from the Grothendieck–Riemann–Roch Theorem. The Mumford Isomorphism allows us to formulate the integrand of the partition function of the Polyakov bosonic string in holomorphic terms. We use the Krichever–Novikov construction of "Algebras of Virasoro Type on Higher Genus Curves" as an example of how to work with the Riemann–Roch theorem. Of course, there are other aspects in the theory of moduli of Riemann surfaces, such as Teichmüller theory, which, because of limited time, could not be covered in the lecture course.

Let me add some remarks on the style. The lecture course and these notes were aimed mainly at physicists, and not so much at mathematicians. But this is not to say that the results are imprecise, rather that there is a mixture between different techniques and different viewpoints. In contrast, a classical mathematical text would stick to one particular aspect, for example, to the analytic theory. There are good reasons for this, but I think that in the context of these lectures we would lose a lot of beautiful relations between the different approaches. Hence these lecture notes might also be useful for students of mathematics who are looking for a short introduction to the various aspects and their interplay. In fact, there were quite a number of mathematics students attending the lectures.

In the lectures, you might find it paradoxical that some comparably simple propositions are proven while there is nothing on the proof of such fundamental theorems as that of Riemann–Roch. But the reason for presenting proofs was not to make a self-contained lecture, but rather to use them as examples of how to work with the defined objects.

In view of p-adic or even adelic string theory, I have included an appendix where I give a short introduction to p-adic numbers. It should be understandable even to someone without prior knowledge of advanced algebra.

Finally, let me thank Prof. Julius Wess for encouraging me to give these lectures and for convincing me that there is a need for a write-up such as this. I would also like to thank the audience for their active participation. Their questions helped me greatly to make the exposition (hopefully) clearer. Let me thank Prof. Rainer Weissauer and Prof. Herbert Popp for valuable suggestions and fruitful discussions concerning the content of Chaps. 7 and 10, Michael Schröder for doing some proof-reading and Günther Schwarz for support in preparing the final printout of the manuscript on the laser printer. I would especially like to thank Dr Ian McArthur for writing an introduction on the application in physics of the mathematical subjects presented and, not least, for correcting my English. However, I accept full responsibility for any remaining errors.

Martin Schlichenmaier (1990)

Contents

Introduction from a Physicist's Viewpoint (by Ian McArthur – 1990)

Advances in the analysis of string theories in recent years have been made using results from the theory of Riemann surfaces and from algebraic geometry, areas of mathematics which may hitherto have been foreign to many physicists, but which any serious student of string theory and more generally conformal field theory probably cannot afford to ignore. This is of course not new in physics, examples being the role of differential geometry in general relativity and group theory and differential geometry in gauge theories.

I have been asked to provide some form of "physical" introduction to these lectures, and I will give my impression of the manner in which some of the concepts developed in the lectures are relevant to the study of two-dimensional conformally invariant field theories and, in particular, bosonic string theory. The list of references is by no means comprehensive, and further details can be obtained from any of the excellent reviews on the subject, for example [1–3].

Two-dimensional Euclidean classical field theory in the presence of a gravitational background is equivalent to the consideration of fields propagating on Riemann surfaces. Amongst the theories with actions which are invariant under diffeomorphisms of the Riemann surface is a special class, namely those for which the action is also invariant under Weyl transformations of the metric ($e^{\Phi}g$ with Φ a smooth function is a Weyl rescaling of the metric g). Such theories are generally termed conformally invariant and are characterized by the vanishing of the energy–momentum tensor. As a result of these invariances, the physics which the action describes depends only on the conformal class of the metric or equivalently the complex structure it determines. Inequivalent theories are parametrized by the moduli, which describe the inequivalent complex structures.

When such a theory is quantized, divergences arise and these must be regulated in order to extract finite quantities from the quantum field theory. In general, it is not possible to choose a regularization scheme which preserves both the diffeomorphism invariance and the Weyl invariance of the classical theory, and the quantum theory is afflicted by an anomaly. The symmetries of the classical action survive in the quantum theory when the field content is

M. Schlichenmaier, Introduction from a Physicist's Viewpoint. In: M. Schlichenmaier, An Introduction to Riemann Surfaces, Algebraic Curves and Moduli Spaces, Theoretical and Mathematical Physics, 1–5 (2007)
DOI 10.1007/978-3-540-71175-9_1

such that the anomalies cancel. This is true for bosonic strings propagating in a flat spacetime of critical dimension. One way to see this is in the Polyakov formulation of string theory. To compute a g-string-loop contribution to the partition function, a functional integral must be performed over the space of metrics on a genus g Riemann surface. The diffeomorphism and Weyl transformations of the theory are gauge transformations, and it is convenient to factor out the volume of the gauge group by a gauge choice. The ghosts which arise in the gauge fixing procedure contribute an anomaly which cancels that of the matter fields in the critical dimension.

Anomaly cancellation ensures the invariance of the quantum theory under Weyl transformations and diffeomorphisms homotopic to the identity. So physical quantities such as the partition function and correlation functions (or Green's functions) are functionals on the space of metrics modulo these groups of transformations – this is Teichmüller space. It can be represented by a slice through the space of metrics transverse to the gauge orbits. In string theory, where an integral over Teichmüller space must be performed (the remnant of the functional integral over all metrics after gauge fixing), different choices of slice correspond to different gauge choices. The independence of the quantum theory in the choice of slice in the absence of anomalies is usually expressed via BRS invariance of the quantization procedure [4, 5].

For Riemann surfaces of genus higher than zero, the diffeomorphism group is not connected, and it is necessary to consider the behaviour of the quantum theory under diffeomorphisms which are not connected to the identity – the equivalence classes of these diffeomorphisms are called modular transformations, and the quotient of Teichmüller space by the group of modular transformations is the moduli space for the Riemann surface. Moduli space is not simply connected and only functions on Teichmüller space that are invariant under modular transformations descend to functions on moduli space – otherwise they determine sections of vector bundles over moduli space. In string theory, modular invariance (which is equivalent to the absence of global gravitational anomalies) is necessary because an integration must be performed over moduli space, and this makes sense only for a well-defined top form on moduli space. It should be noted that the locations of the insertions of field operators in correlation functions are modular parameters for a punctured Riemann surface, the moduli space coming from a quotient of the space of metrics by the group of diffeomorphisms preserving the locations of the punctures.

Moduli space is a complex space. Belavin and Knizhnik [6] showed that for closed bosonic string theory, the integrand in the integral over moduli space in the partition function can be arranged such that it is locally the squared modulus of a holomorphic function on moduli space – this has been termed holomorphic factorization and is closely related to some results in algebraic geometry. The determinants which arise from the functional integrals over ghost fields and spacetime fields (for a flat spacetime) can be interpreted as the norms of holomorphic sections of the line bundles H and E^{-13}, where H

is the canonical bundle of moduli space and E is the highest exterior power of the vector bundle of abelian differentials. The calculation of Belavin and Knizhnik shows that the first Chern class of the bundle $H \otimes E^{-13}$ vanishes. In fact, the bundle is trivial, a result established by Mumford some time ago [7]. Using this result, it is possible to choose holomorphic sections of H and E in such a way that the partition function is given entirely by the integration measure on moduli space, the integrand from the functional determinants contributing only a constant. Single valuedness of the integration measure on moduli space allows identification of factors in it with modular forms.

Moduli space is also noncompact, and divergences in string theory arise from integration over asymptotic regions of moduli space. For example, in bosonic string theory, divergences in the partition function can be interpreted as being due to integrals of the form $\int dL \exp(-E_0 L)$ corresponding to the propagation of a tachyon with energy $E_0 < 0$ along a long thin tube of length L. The integral diverges as a result of contributions from the asymptotic region $L \to \infty$. The nature of these divergences can be analysed more geometrically by adding points (in general infinitely many) to moduli space which compactify it while preserving the complex structure – much as the complex plane can be compactified to the Riemann sphere by adding a point at infinity. The added points correspond to pinched Riemann surfaces. The measure for integration over moduli space is constructed from a meromorphic section of the line bundle $H \otimes E^{-13}$ over the compactified moduli space, whose divisor can be interpreted in terms of the expected divergences due to the tachyon.

With respect to string theory, the discussion so far has centred on the Polyakov or functional integral formulation of the theory. There also exists a more "local" formulation of Weyl invariant (or conformal) field theories. In local complex coordinates z on a Riemann surface, the metric g has the form $\varphi(z, \bar{z}) \, dz \otimes d\bar{z}$. The scale factor φ decouples from a Weyl invariant theory, and all relevant information about the metric is contained in the local complex coordinates and their transition functions. Weyl invariance is reflected as an invariance under local changes of complex coordinates – these are termed conformal transformations. The fields in a conformal field theory (including composite fields such as the energy–momentum tensor) are sections of powers of the canonical bundle K, and when expressed relative to local holomorphic trivializations dz of K, they transform by powers of $\partial z' / \partial z$ under local conformal transformations $z \to z'(z)$. Such a theory can be quantized within a local coordinate chart in a manner analogous to canonical quantization, and the problem is to patch this local information together on a higher genus Riemann surface [8]. Recently, several promising approaches have emerged. The first [9–11] is based on consideration of a Riemann surface with a disc removed. A vacuum state can be associated with the boundary of the Riemann surface, and it encodes all the information about the Riemann surface relevant to the conformal field theory. Correlation functions on the Riemann surface are computed by inserting field operators between this vacuum and a vacuum state associated with the boundary of the disc. This is closely related to work by mathematicians on infinite-dimensional Grassmann manifolds and solutions of the KP equation [12, 13]. Information about the moduli space of the Riemann surface enters through meromorphic sections of line bundles on the Riemann surface which are holomorphic outside the disc. The Weierstrass gap theorem is an essential component in this analysis.

The second approach requires the consideration of a twice punctured Riemann surface [14]. A special choice of basis (called the Krichever–Novikov basis) for meromorphic sections of line bundles, in which all poles lie at the punctures, allows the decomposition of field operators (which are sections of the bundles) into modes analogous to the standard decomposition on the sphere. Many of the calculational techniques used on the sphere can be reproduced for higher genus surfaces in this basis. Information about the moduli space is contained in the Krichever–Novikov basis.

This is only a brief and by no means complete list of the points of contact between the content of the lectures and physics. Important omissions are the work in [15–17].

To conclude, I should include the warning that mathematics is one of the tools of a physicist's trade, but the tool and the product should not be confused. The product should be a description of observed phenomena. Whether string theory will meet this criterion is yet to be seen. Whatever the outcome, I wish the reader as much enjoyment as Martin Schlichenmaier provided his audience in his presentation of the lectures.

References

1. L. Alvarez-Gaumé and P. Nelson, Riemann Surfaces and String Theories (in) *Supersymmetry, Supergravity and Superstrings 86*, World Scientific, Singapore 1986.
2. P. Nelson, Lectures on Strings and Moduli Space, *Phys. Rep.* **149** (1987) 304.
3. E. D'Hoker and D. Phong, The Geometry of String Pertubation Theory, to appear in *Rev. Mod. Phys.*
4. D. Friedan, E. Martinec and S. Shenker, *Nucl.Phys* **B271** (1986) 93.
5. S.B. Giddings and E. Martinec, *Nucl. Phys.* **B281** (1986) 91.
6. A. Belavin and V. Knizhnik, *Sov.Phys. JETP* **64** (1987) 214.
7. D. Mumford, *L'Enseign. Math.* **23** (1977) 39.
8. E. Martinec, *Superstrings '87*, World Scientific, Singapore 1987.
9. N. Ishibashi, Y. Matsuo and H. Ooguri, *Mod. Phys. Lett.* **A2** (1987) 119.
10. L. Alvarez-Gaumé, C. Gomez and C. Reina, *Superstrings '87*, World Scientific, Singapore 1987.
11. E.W.Witten, *Commun. Math. Phys* **113** (1988) 529.

12. E. Date, M. Jimbo, M. Kashiwara and T.Miwa, *Proceedings of RIMS Symposium on Nonlinear Integrable Systems*, World Scientific, Singapore, 1983.
13. G. Segal and G. Wilson, *Publ. Math. IHES* **61** (1985) 1.
14. I.M. Krichever and S.P. Novikov, *Funkts. Anal. Prilozh.* **21(2)** (1987) 46; **21(4)** (1987) 47.
15. H. Sonoda, *Phys. Lett.* **178B** (1986) 390.
16. E. Verlinde and H. Verlinde, *Nucl.Phys.* **B288** (1987) 357.
17. L. Alvarez-Gaumé, J. Bost, G. Moore, P. Nelson and C. Vafa, *Commun. Math. Phys.* **112** (1987) 503.

1

Manifolds

1.1 Generalities

First we have to understand what manifolds are. Our rough idea is that a manifold should look locally like $\mathbb{R}^n$, which we want to make precise in the following. Let M be a topological space (essentially a set with a rule for which subsets we are considering as open sets) and $\{U_i, i \in J\}$ a family of open sets in M with

$$\bigcup_{i \in J} U_i = M \ . \tag{1.1}$$

For every U_i there should be a map

$$\varphi_i : U_i \to W_i \subset \mathbb{R}^n$$

for a fixed $n \in \mathbb{N}$. The W_i are assumed to be open and the φ_i are topological isomorphisms. In other words we require φ_i to be bijective and φ_i and φ_i^{-1} to be continuous. (U_i, φ_i) is called a *coordinate patch* or sometimes just a coordinate. The *dimension* of M is defined as the number n. The collection of such coordinate patches is called an *atlas*.

Unfortunately this definition is a little bit too general. It allows examples which we would not intuitively call a manifold. To get rid of them we require two additional things:

(1) M has to be a *separated* topological space. This means that every two distinct points have disjoint open neighbourhoods. Another name for this is that M is a *Haussdorff space*.
(2) M has to be *paracompact*.

If nothing else is said we assume that our manifolds M are connected. This says that given two points p, q on our manifold there always exists a continuous path from p to q, i.e. a continuous map $\alpha : [0, 1] \to M$ such that $\alpha(0) = p$ and $\alpha(1) = q$. By the connectedness the dimension of M is well defined.

M. Schlichenmaier, Manifolds. In: M. Schlichenmaier, An Introduction to Riemann Surfaces, Algebraic Curves and Moduli Spaces, Theoretical and Mathematical Physics, 7–15 (2007)
DOI 10.1007/978-3-540-71175-9_2   © Springer-Verlag Berlin Heidelberg 2007

Definition 1.1. *A separated topological space M is called paracompact if every open covering of M allows a locally finite refinement.*

Instead of defining the above notion in detail we state the following:

Theorem 1.2. *Let M be a separated topological space, then the following two conditions are equivalent:*

(1) M is paracompact.
(2) For every open covering of M there exists a subordinated partition of unity.

What do we understand by such a *partition of unity?* Let

$$\mathcal{U} = (U_i)_{i \in J}$$

be a covering of M by open sets. Then a subordinated partition of unity is given by a family of continuous functions

$$(\tau_i)_{i \in J} : M \to [0, 1] \tag{1.2}$$

with

(a) $(\tau_i)_{i \in J}$ is locally finite, which says for every point $x \in M$ there is a neighbourhood W of x with

$$\tau_i|_W \equiv 0$$

except for a finite number of $i \in J$,
(b) for all $x \in M$ we have

$$\sum_{i \in J} \tau_i(x) = 1,$$

(c)

$$\operatorname{supp} \tau_i := \overline{\{x \in M \mid \tau_i(x) \neq 0\}} \subset U_i.$$

Later we will see this is a rather useful property to define local objects and glue them together to global objects.

In the following we are mainly interested in compact manifolds. A manifold is called *compact* if every open covering of M contains a finite subcovering, i.e. given $\{U_i \mid i \in J\}$ with (1.1) then there exist $i_1, i_2, \ldots, i_k \in J$ with $\bigcup_{l=1,\ldots,k} U_{i_l} = M$. Hence compact manifolds are obviously paracompact.

Let us take a closer look at the coordinate patches (see Fig. 1.1). Let U_i, U_j be two different neighbourhoods with $U_i \cap U_j \neq \emptyset$.

The points in $U_i \cap U_j$ lie in two coordinate patches. Hence we can define

$$\psi_{ji} := \varphi_j \varphi_i^{-1} : \varphi_i(U_i \cap U_j) \to \varphi_j(U_i \cap U_j)$$

as a map from a subset of $\mathbb{R}^n$ to $\mathbb{R}^n$. The map ψ_{ji} is called a *coordinate change map* or a *glueing function*. Obviously it is continuous and bijective.

We might require additional properties for these ψ_{ji} and get restricted sets of manifolds. We call $(M, \mathcal{U})$ a *differentiable manifold* if all glueing functions are differentiable functions. (If we use the word differentiable function we always assume infinitely often differentiable, i.e. a C^∞ function.) It is called a *real analytic manifold* if the glueing functions are analytic functions.

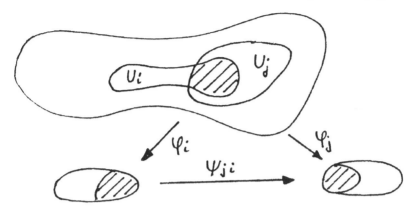

Fig. 1.1. Overlap of coordinates

Example 1.3. The *sphere* S^n is given as

$$S^n \quad := \quad \left\{ x = (x_1, x_2, \cdots, x_{n+1}) \in \mathbb{R}^{n+1} \;\middle|\; ||x||^2 := \sum_{i=1}^{n+1} x_i^2 = 1 \right\}$$

As coordinate patches we use the half-spheres

$$U_{kj} := \{ x \in S^n \mid (-1)^j x_k > 0 \}, \quad j := 1, 2, \; k := 1, 2, \ldots, n+1.$$

The coordinate functions are defined as

$$h_{kj} : U_{kj} \to D^n := \{ x \in \mathbb{R}^n \mid ||x|| < 1 \},$$
$$h_{kj}(x_1, x_2, \ldots, x_{k-1}, x_k, x_{k+1}, \ldots, x_n) = (x_1, x_2, \ldots, x_{k-1}, x_{k+1}, \ldots, x_n).$$

This is nothing else than dropping the kth position, hence flattening the half-spheres. The inverse map is given by inserting at the kth position

$$(-1)^j \cdot \sqrt{\left(1 - \sum_{i \neq k} x_i^2 \right)}$$

and shifting all entries with indices $p \geq k$. Obviously the map

$$h_{kj}(h_{lm})^{-1}$$

is differentiable (even real analytic) where it is defined. This covering of the sphere is called the *standard covering*.

1.2 Complex Manifolds

In the above definitions we leave everything the same except replacing $\mathbb{R}^n$ by $\mathbb{C}^n$. Now

$$\psi_{ji} : U \subset \mathbb{C}^n \to W \subset \mathbb{C}^n$$

and we require these to be holomorphic functions. In detail, if $(z_1, z_2, \ldots, z_n)$ are coordinates in U and $(w_1, w_2, \ldots, w_n)$ are coordinates in W, then

$$w_j = w_j(z_1, z_2, \ldots, z_n), \quad j = 1, \ldots, n .$$

We require w_j to be holomorphic functions of z_i and the holomorphic functional determinant

$$\frac{\partial(w_1, w_2, \ldots, w_n)}{\partial(z_1, z_2, \ldots, z_n)} \neq 0.$$

If these conditions are fulfilled we call M a *complex manifold* of (complex) dimension n.

If we identify

$$\mathbb{C}^n \to \mathbb{R}^{2n}, \quad z_i = x_i + iy_i \mapsto (x_i, y_i),$$

M also has a structure as a real $2n$-dimensional differentiable manifold. Vice versa we can start with a differentiable $2n$-dimensional manifold where there exist real coordinates which we can group in pairs (x_i, y_i) and (g_j, h_j) such that

$$z_i = x_i + iy_i, \quad w_j = g_j + ih_j$$

are holomorphic coordinates in the above sense. This can be written as the requirement that they obey the *Cauchy–Riemann differential equations*

$$\frac{\partial g_j}{\partial x_i} = \frac{\partial h_j}{\partial y_i}, \quad \frac{\partial h_j}{\partial x_i} = -\frac{\partial g_j}{\partial y_i}, \quad i, j = 1, 2, \ldots, n.$$

Example 1.4. The *complex projective space*

$$\mathbb{P}^n := (\mathbb{C}^{n+1} \setminus \{0\})/ \sim .$$

We define two elements $z, w \in \mathbb{C}^{n+1}$ as equivalent under the relation $\sim$ if $z = \lambda \cdot w$ with $\lambda \in \mathbb{C}, \lambda \neq 0$. In this way we get a map

$$\pi : \mathbb{C}^{n+1} \setminus \{0\} \to \mathbb{P}^n, \quad (z_0, z_1, \cdots, z_n) \mapsto (z_0 : z_1 : \cdots : z_n).$$

The coordinates z_i are called *homogeneous coordinates*. Of course they are only unique up to a constant scalar multiple and hence not coordinates on the manifold in the above sense. The topology on $\mathbb{P}^n$ is defined as the topology coming from π. In particular π is required to be continuous. Moreover W is open in $\mathbb{P}^n$ if and only if $\pi^{-1}(W)$ is open in $\mathbb{C}^{n+1} - \{0\}$.

We also have a system of standard coordinates (U_i, φ_i) to define a manifold structure on the projective space:

$$U_i := \{(z_0 : z_1 : \cdots : z_n) \mid z_i \neq 0\}, \quad i = 0, 1, \ldots, n \tag{1.3}$$

and

$$\varphi_i : U_i \to \mathbb{C}^n, \qquad (z_0 : z_1 : \cdots : z_n) \mapsto \left(\frac{z_0}{z_i}, \frac{z_1}{z_i}, \ldots, \frac{z_{i-1}}{z_i}, \frac{z_{i+1}}{z_i}, \ldots, \frac{z_n}{z_i} \right).$$

With this map U_i is even identified with the whole space $\mathbb{C}^n$. The complement $\mathbb{P}^n \setminus U_i$ is a $n-1$ dimensional manifold, the hyperplane "lying at infinity". In this sense $\mathbb{P}^n$ is a compactification of the affine space $\mathbb{C}^n$. The glueing functions are given by (we assume $j < i$ in this example)

$$\psi_{ji} : \mathbb{C}^n \setminus \{w_j \neq 0\} \to \mathbb{C}^n \setminus \{w'_{i-1} \neq 0\} :$$

$$(w_0, w_1, \ldots, w_j, \ldots, w_i, \ldots, w_{n-1}) \mapsto$$

$$(w_0 : w_1 : \ldots, w_j : \ldots : w_{i-1} : 1 : w_i : \ldots : w_{n-1}) =$$

$$\left(\frac{w_0}{w_j} : \frac{w_1}{w_j} : \ldots : 1 : \ldots : \frac{w_{i-1}}{w_j} : \frac{1}{w_j} : \frac{w_i}{w_j} : \ldots : \frac{w_{n-1}}{w_j} \right) \mapsto$$

$$\left(\frac{w_0}{w_j}, \frac{w_1}{w_j}, \ldots, \frac{w_{j-1}}{w_j}, \frac{w_{j+1}}{w_j}, \ldots, \frac{w_{i-1}}{w_j}, \frac{1}{w_j}, \frac{w_i}{w_j}, \ldots, \frac{w_{n-1}}{w_j} \right).$$

Obviously, this is a holomorphic map.

In the special case $n = 1$ we get the *Riemann sphere* with the standard covering

$$U_0 := \{(z_0 : z_1) \mid z_0 \neq 0\}$$
$$U_1 := \{(z_0 : z_1) \mid z_1 \neq 0\}.$$

We write

$$U_1 := \{(w : 1) \mid w \in \mathbb{C}\} \cong \mathbb{C}$$

and obtain as point "∞"

$$\mathbb{P}^1 \setminus U_1 = \{(1 : 0)\}.$$

We also need the notion *orientable manifold*. We call a manifold orientable if it admits a covering with coordinate patches where all coordinate change maps (the real ones) respect the orientation. In the case of differentiable manifolds this is equivalent to the requirement that the functional determinant of the ψ_{ij} is greater than zero.

Remark 1.5. (only for those readers who are familiar with differential forms) Orientability is essentially the fixing of one top dimensional differential form as positive. Let $x_1, x_2, \ldots, x_n$ be local coordinates, $y_1, y_2, \ldots, y_n$ the images of these under a coordinate transformation then

$$dy_1 \wedge dy_2 \cdots \wedge dy_n = \left(\sum_i \frac{\partial y_1}{\partial x_i} dx_i \right) \wedge \left(\sum_i \frac{\partial y_2}{\partial x_i} dx_i \right) \cdots \wedge \left(\sum_i \frac{\partial y_n}{\partial x_i} dx_i \right)$$

$$= \det \left(\frac{\partial y_j}{\partial x_i} \right) dx_1 \wedge dx_2 \cdots \wedge dx_n$$

which proves the equivalence.

Proposition 1.6. *Complex manifolds are always orientable.*

Proof. Let

$$J_{hol} = \det\left(\frac{\partial w_i}{\partial z_j}\right)$$

be the holomorphic functional determinant. For checking the orientability we need the real functional determinant which is given by

$$J = \det\begin{pmatrix} \dfrac{\partial w_i}{\partial z_j} & \dfrac{\partial w_i}{\partial \overline{z}_j} \\ \dfrac{\partial \overline{w}_i}{\partial z_j} & \dfrac{\partial \overline{w}_i}{\partial \overline{z}_j} \end{pmatrix}.$$

Now due to the holomorphic structure

$$\frac{\partial w_i}{\partial \overline{z}_j} = \frac{\partial \overline{w}_i}{\partial z_j} = 0,$$

hence

$$J = \det\left(\frac{\partial w_i}{\partial z_j}\right) \cdot \det\left(\frac{\partial \overline{w}_i}{\partial \overline{z}_j}\right) = \det\left(\frac{\partial w_i}{\partial z_j}\right) \cdot \overline{\det\left(\frac{\partial w_i}{\partial z_j}\right)}$$

$$= |J_{hol}|^2 > 0. \quad \square$$

Proposition 1.7. *A two-dimensional orientable and compact differentiable manifold always admits a complex structure.*

Proof. This is only a sketch for the reader who is familiar with the notion of metrics on a manifold. If not, just skip it. Let X be the two-dimensional orientable manifold. Locally we can always define a metric on X. Such an object is a differentiable section of the bundle $\mathbf{T}^*(M) \otimes \mathbf{T}^*(M)$ satisfying the conditions of a scalar product on $\mathbf{T}(M)$. By partition of unity (1.2) we can glue the local data into a global section g. Now it is possible (in dimension 2 !) to find a set of oriented coordinates x, y compatible with the differentiable structure such that g is given by

$$g = e^f (dx \otimes dx + dy \otimes dy),$$

where f is a differentiable function. These coordinates are called *isothermal coordinates*. With these coordinates $z = x + iy$ are holomorphic coordinates because the coordinate change from one isothermal coordinate system to another is holomorphic. This follows from a not difficult but tedious calculation.

Now we are able to define the central objects of the lecture.

Definition 1.8. *A one-dimensional connected complex manifold is called a Riemann surface.*

We saw above that a compact Riemann surface is from the viewpoint of the real numbers a two-dimensional compact, connected and orientable differentiable manifold. Because we are mostly concerned with compact Riemann surfaces we will use for simplicity the notation "Riemann surface" to denote a compact Riemann surface.

In certain circumstances it might be useful to consider noncompact Riemann surfaces also or even Riemann surfaces with boundaries. If we use them later on we will specify it explicitly.

1.3 The Classification Problem

We want to consider two manifolds M and M' as the same manifold if there is a bijection between them which respects the structure under consideration. Of course this does not depend on the sets M and M' alone but on their topology and even on the set of coordinates chosen.

From the *topological viewpoint* we say $M \cong M'$ if there is a bijection

$$\psi : M \rightarrow M'$$

which is continuous in both directions. If this is the case we say ψ is a *topological isomorphism* or a *homeomorphism*. From the topological viewpoint there is no need to consider coordinate patches. In the case of compact, orientable two-dimensional manifolds there is for every integer ≥ 0 just one isomorphy class. This number which classifies the manifolds from the topological viewpoint is called *genus*. It will be defined in Chap. 2. The isomorphy class of a manifold is also called its topological type.

If we consider the *differentiable structure* we have to require of the topological isomorphism that it respects this structure. In detail, if

$$(M, \mathcal{U}) \quad \text{and} \quad (M', \mathcal{U}')$$

are two differentiable manifolds and

$$f : M \rightarrow M'$$

is a topological isomorphism then we call f a *differentiable isomorphism* or a *diffeomorphism* if the following is valid. For every $a \in M$ let $b = f(a) \in M'$ be the image point. We choose coordinate patches around a and b:

$$(U, \phi), \quad a \in U, \quad (V, \psi), \quad b \in V.$$

Now we build the map

$$f_{VU} := \psi \circ f \circ \phi^{-1}.$$

This map is defined on the set

$$\phi(f^{-1}(V) \cap U)$$

in $\mathbb{R}^n$ and takes values in the set

$$\psi(V) \subset \mathbb{R}^n.$$

It is just a normal function with arguments and values in $\mathbb{R}^n$. We call f a differentiable map if f_{UV} is differentiable for all possible choices made. We say f is a diffeomorphism if f and f^{-1} are differentiable maps.

In general, every topological type splits into different diffeomorphy types (classes of nondiffeomorphic manifolds). But in the case of two-dimensional, compact and orientable manifolds there is just one diffeomorphy type for every genus. In other words, if there is a topological isomorphism between two of them there is also a diffeomorphism between them. This is not the case in higher dimensions. The first example was discovered by Milnor. The topological S^7 admits 28 different differentiable structures. This should warn us that a differentiable manifold is not just given as a topological space, it always has to come with a set of coordinates. There have been some rather spectacular results in the last few years. For example, it was proven some time ago that $\mathbb{R}^n$ for $n \neq 4$ has a unique differentiable structure. The missing case withstood solution for a long time. By a chain of very deep mathematics it was proven a few years ago that in fact $\mathbb{R}^4$ has a continuous family of nonisomorphic differentiable structures. An important piece in the chain was the use of the mathematical version of Yang–Mills theory. It was possible to show that even for compact four-dimensional manifolds the topological type does not fix the diffeomorphy type.

In the case of complex manifolds we can define, analogous to above, what we mean by an analytic map. For $f : M \rightarrow M'$ to be an analytic map, we require that the maps f_{UV}, which are now maps from a subset of $\mathbb{C}^n$ to $\mathbb{C}^n$, are holomorphic. We call f an *analytic isomorphism* (sometimes also called a *biholomorphic map*) if f is bijective and f and f^{-1} are holomorphic.

As above we can ask how many different analytic types lie over a fixed topological type. In the case of Riemann surfaces we get for every type $g \geq 1$ a "continuous family" of different types depending on a certain number of complex parameters. Let us note the result:

$g = 0$ there is only one analytic type,
$g = 1$ a family with 1 complex parameter,
$g \geq 2$ a family with $3g - 3$ complex parameters.

The precise meaning will be given in Chap. 7.

Hints for Further Reading

On the generalities of manifolds see textbooks on advanced analysis. Let me only mention:

[BJ]   Bröcker, Th., Jänich, K., *Einführung in die Differentialtopologie*, Springer, 1973.

[Wa]   Warner, F., *Foundations of Differentiable Manifolds and Lie Groups*, Springer, 1983.

In our context only the paragraphs 1 to 4 of [BJ] are of importance. The book gives a short but quite understandable introduction to manifolds and vector bundles. The book [Wa] is more advanced.

For an introduction to complex manifolds, see

[CH]   Chern, S.S., *Complex Manifolds Without Potential Theory*, Springer, 1979.

It is rather short but contains the important definitions we need and much more material.

2

Topology of Riemann Surfaces

2.1 Fundamental Group

Let us fix a Riemann surface M. The first topological characteristic of M is essentially the set of "different" loops on the surface. For this we want to identify two loops if we can transform the first by a continuous deformation into the second. Let us make this more precise. We denote by $I = [0,1]$ the unit interval in the real numbers. A *path* α from the point p to the point q on the manifold M is a continuous map

$$\alpha : I \to M, \qquad \alpha(0) = p, \ \alpha(1) = q.$$

If

$$\alpha(1) = \alpha(0),$$

the path is called a *loop*. If β is another path from p to q then we call α homotopic to β ($\alpha \sim \beta$) if there exists a homotopy between the two. Intuitively this means we can deform α into β. In exact terms a *homotopy* H is defined as a map

$$H : I \times I \to M$$

with

$$h(t,0) = \alpha(t), \quad h(t,1) = \beta(t)$$

for all $t \in I$ and

$$h(0,u) = p, \quad h(1,u) = q$$

for all $u \in I$. The paths h_u defined by

$$h_u = h(.,u) : I \to M$$

are the paths "between" α and β (see Fig. 2.1).
If g is a reparametrization of I which respects the orientation

$$g : I \to I, \qquad g(0) = 0, \ g(1) = 1.$$

M. Schlichenmaier, Topology of Riemann Surfaces. In: M. Schlichenmaier, An Introduction to Riemann Surfaces, Algebraic Curves and Moduli Spaces, Theoretical and Mathematical Physics, 17–30 (2007)
DOI 10.1007/978-3-540-71175-9_3

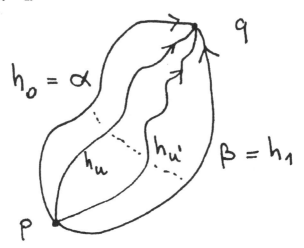

Fig. 2.1. A family of homotopic paths

then $\beta = \alpha \circ g$ will not be the same path as α. But at least $\beta \sim \alpha$. A homotopy in this case is given by

$$h(t, u) = \alpha(t + u \cdot (g(t) - t)).$$

Homotopy is an equivalence relation. Let us denote the homotopy class of a path α as $[\alpha]$.

We can compose paths. Let α be a path from p to q and β be a path from q to v, then by $\gamma = \alpha\beta$ we mean the path from p to v defined by

$$\gamma(t) = \begin{cases} \alpha(2t), & 0 \le t \le \frac{1}{2} \\ \beta(2t - 1), & \frac{1}{2} \le t \le 1 . \end{cases}$$

This composition is not associative. We can also define an inverse path α^{-1} for α. It is defined by

$$\alpha^{-1}(t) = \alpha(1 - t).$$

If we carry out the above operations with paths α' and β' which are homotopic to α, resp. β, then the result $\alpha'\beta'$, resp. α'^{-1}, will be homotopic to $\alpha\beta$, resp. α^{-1}. So the above operations are well defined on the homotopy classes of paths. In fact here the composition is associative. If we fix a point $p \in M$ and restrict ourselves to loops starting and ending at the point p we can compose without any restriction. We calculate $\alpha\alpha^{-1} \sim 1_p$, where 1_p is the constant loop

$$I \to M, \qquad 1_p(t) = p .$$

Definition 2.1. *The fundamental group of M with respect to the point p is given by*

$$\pi(M, p) := \{ all\ homotopy\ classes\ of\ loops\ at\ the\ point\ p \}$$

with the above composition as multiplication.

Proposition 2.2. $\pi(M, p)$ *is a group. The constant loop* $[1_p]$ *is the unit element in the group. If we choose another point* q *then*

$$\pi(M, p) \cong \pi(M, q).$$

If γ *is a path from* p *to* q, $[\gamma]$ *its homotopy class, then the isomorphism is given by*

$$[\alpha] \mapsto [\gamma^{-1}][\alpha][\gamma].$$

Important for the last result is that our Riemann surfaces are connected. In the following we leave out the base point p and just write $\pi(M)$.

Definition 2.3. M *is called simply connected if*

$$\pi(M) = \{1\}.$$

Intuitively, this says that all loops are contractible to one point.

Example 2.4. The sphere S^2 (Fig. 2.2).
On the sphere all loops are contractible to the constant loop, hence it is simply connected.

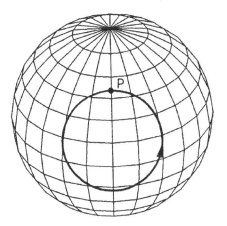

Fig. 2.2. The sphere

Example 2.5. The torus T (Fig. 2.3).
The loops α and β are not contractible. For every other loop γ the class can be written as

$$[\gamma] = [\alpha]^n [\beta]^m, \qquad n, m \in \mathbb{Z}.$$

$\pi(T)$ is an abelian group. This can be seen by cutting the torus at the paths α and β and flattening it into a plane. We get the situation drawn in Fig. 2.4. If we set $\gamma_1 = \beta\alpha$ and $\gamma_2 = \alpha\beta$ we see $\gamma_1 \sim \gamma_2$; hence

$$[\alpha][\beta] = [\beta][\alpha].$$

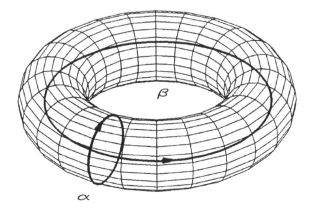

Fig. 2.3. The torus

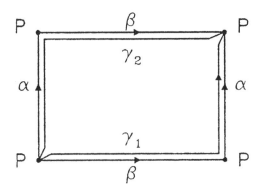

Fig. 2.4. The torus situation

Example 2.6. M_2 (see Fig. 2.5).
In the figure $\alpha_1, \alpha_2, \beta_1, \beta_2$ are again noncontractible loops and $\pi(M)$ is generated by them. But now there is no reason why we should have

$$[\beta_1][\alpha_1] = [\alpha_1][\beta_1].$$

In fact, this is not valid. The fundamental group for this manifold is non-abelian. If we cut our surface along the above loops we get the situation drawn in Fig. 2.6.
In the above notation α_i^{-1} has as image the same set on M_2 as α_i but with different orientation. We see that we can contract the boundary of this polygon on the manifold. Hence we know at least that

$$\alpha_1\beta_1\alpha_1^{-1}\beta_1^{-1}\alpha_2\beta_2\alpha_2^{-1}\beta_2^{-1} \sim 1.$$

In fact this is the only relation in $\pi(M_2)$.

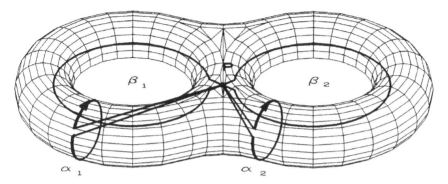

Fig. 2.5. Higher genus

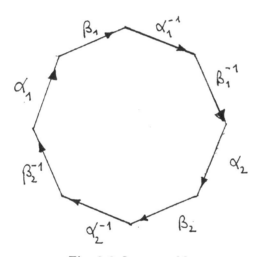

Fig. 2.6. Loops on M_2

2.2 Simplicial Homology

We start with a triangulation of our Riemann surface M. By a *triangulation* we mean a covering of M by "deformed" triangles such that any two triangles have either no intersection at all or have a vertex or a complete edge in common. For edges we require that they have at most one vertex in common. Figure 2.7 shows such a triangulation.

Here we will accept that our Riemann surfaces admit triangulations. The triangles are also called *2-simplices*, the edges *1-simplices* and the vertices *0-simplices*.

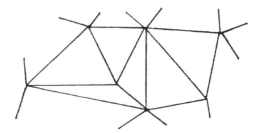

Fig. 2.7. Triangulation

Let $p_1, p_2, \ldots, p_k$ be the 0-simplices. The 1-simplices can be given by $< p_r p_s >$, the 2-simplices by $< p_r p_s p_t >$. The simplex as purely geometric object does not depend on the order of the points chosen to describe it. So we can always assume after a permutation $r < s(< t)$ and so on. By the *orientation* of the simplex we mean the sign of the required permutation to bring it in the above form. In this sense

$$< p_2 p_1 >= - < p_1 p_2 >, \qquad < p_3 p_2 p_1 >= - < p_1 p_2 p_3 > .$$

The group of *n-chains* C_n is the free abelian group generated by the set of positively oriented n-simplices. Hence

$$C_n = \bigoplus_{i=1,\ldots,l} \mathbb{Z}\,\sigma_i,$$

where $\sigma_i, i = 1, 2, \ldots, l$ are all positively oriented n-simplices. Roughly speaking a chain is a collection of simplices with integer multiplicities.

If we take the triangle $< p_1 p_2 p_3 >$ then it has a boundary consisting of edges

$$< p_1 p_2 > + < p_2 p_3 > + < p_3 p_1 > = < p_1 p_2 > + < p_2 p_3 > - < p_1 p_3 > .$$

Figure. 2.8 shows this boundary.

By this analogy we define a boundary map

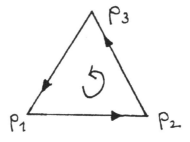

Fig. 2.8. The triangle and its boundary components

$$\delta_n : \ C_n \ \to \ C_{n-1}$$

by setting

$$\delta_0 < p_r >= 0, \qquad \delta_1 < p_r p_s >= p_s - p_r,$$
$$\delta_2 < p_r p_s p_t >=< p_r p_s > + < p_s p_t > - < p_r p_t >$$

and $\delta_k = 0$ for other values of k. Of course, in the above we extend the map on the whole of C_n by linear continuation.

Proposition 2.7.

$$\delta_{n-1} \delta_n = 0.$$

Proof. It is only necessary to check

$$\delta_1 \delta_2 < p_r p_s p_t > = \delta(< p_r p_s > + < p_s p_t > - < p_r p_t >)$$
$$= p_s - p_r + p_t - p_s - p_t + p_r = 0. \qquad \square$$

We call a n-chain c *closed* if $\delta_n(c) = 0$. We define the n-*cycles* Z_n as the n-chains which are closed and the n-*boundaries* B_n as the n-chains which are boundaries of $(n+1)$-chains:

$$Z_n := \ker \delta_n := \{c \in C_n \mid \delta_n(c) = 0\},$$
$$B_n := \delta_{n+1}(C_{n+1}).$$

By the above proposition we see

$$B_n \subseteq Z_n$$

and hence we can make the following

Definition 2.8. *The factor group*

$$H_n = \frac{Z_n}{B_n}$$

is called the nth simplicial homology group.

Let us calculate the easier parts first.

H_0: Because $\delta_0 = 0$ every 0-simplex is a cycle. Now every two points p_1 and p_k can be connected by a path of edges

$$< p_1 p_2 > \cup < p_2 p_3 > \cup \cdots \cup < p_{k-1} p_k > .$$

We calculate

$$p_k - p_1 = \delta_1 < p_1 p_2 > + \delta_1 < p_2 p_3 > + \cdots + \delta_1 < p_{k-1} p_k > .$$

Hence p_k and p_1 just differ by a boundary of 1-chain. In other words, they represent the same homology class. Thus after choosing a point $p \in M$

$$H_0(M) = \mathbb{Z} \cdot <p> .$$

H_2: Here no 2-boundaries can occur because $\delta_3 = 0$. Hence H_2 contains just the 2-chains which have no boundary. Now a chain has exactly no boundary if it is a multiple of the whole manifold. If $\sigma_1, \sigma_2, \ldots, \sigma_l$ are all 2-simplices then

$$H_2(M) = \mathbb{Z} \cdot \left(\sum_{i=1}^{l} \sigma_i \right).$$

Earlier we have used for the chain and its homology class the same notation for simplicity. We will do this further on if there is no ambiguity.

H_1: Up to now nothing interesting has happened, but this will change.

Proposition 2.9.
$$H_1(M) \cong \pi(M)^{\mathrm{ab}}.$$

If we have a group G then the group G^{ab} is the *abelianization* of G. It is defined by dividing out the so-called commutator subgroup which is the normal subgroup generated by the commutators

$$aba^{-1}b^{-1}, \qquad a, b \in G.$$

If the group is given by generators and relations we can get G^{ab} by adding the additional relations $ab = ba$ for every pair of generators. In view of the examples of the last section we know

$$H_1(S^2) = 0, \quad H_1(T) \cong \mathbb{Z} \oplus \mathbb{Z}, \quad H_1(M_2) = \mathbb{Z} \oplus \mathbb{Z} \oplus \mathbb{Z} \oplus \mathbb{Z}.$$

The generators are given by α and β. We will see this again in the following.
    Let us make first the following:

Definition 2.10. *The rank of the group* $H_i(M)$ *is called the ith Betti number* b_i.

We know $b_0 = b_2 = 1$ and we are interested in what b_1 is. To calculate it we transform our surface M into the so-called *topological normal form*. Essentially it is only a manipulation with the triangulation without changing the topological type. For this triangulation we also allow arbitrary polygons as elements. By certain glueing and pasting[1] of the triangulation we get as the final result either the picture of S^2 or just one polygon with $4g$ vertices. The $4g$ vertices correspond to one point on the manifold; the $4g$ edges of the polygon correspond to $2g$ edges on the manifold because it is reconstructed by glueing pairs of edges together. The number g is called the *topological genus of M*. In the case of S^2 we set $g = 0$. It will turn out that $b_1 = 2g$.
But first let us consider some examples for the glueing, see Fig 2.9, 2.10, 2.11.

---

[1] See Seifert/Threlfall, [ST], pp. 135–142.

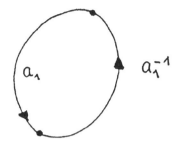

Fig. 2.9. Identification for the sphere

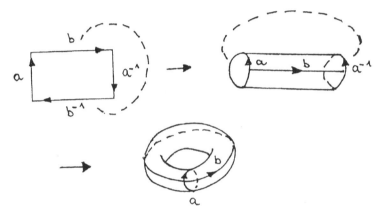

Fig. 2.10. Identification for the torus

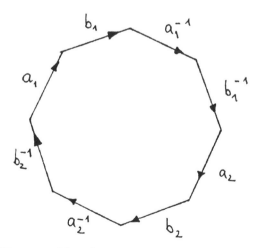

Fig. 2.11. Identification for a manifold of genus 2

The general picture is that by increasing the genus by 1 another group of four edges $a_k b_k a_k^{-1} b_k^{-1}$ shows up. By glueing together a $4g$-gon we get the sphere with g handles attached. This is the so-called handle model of the manifold. So in fact the genus gives the number of "holes" in a Riemann surface embedded in $\mathbb{R}^3$.

Unchanged by the allowed manipulation of the triangulation is the *Euler characteristic*

$$\chi(M) := \alpha_0 - \alpha_1 + \alpha_2,$$

where α_i is the number of i-simplices in the triangulation. From the final polygon we calculate for a manifold M with $g(M) > 0$

$$\chi(M) = 1 - 2g + 1 = 2 - 2g(M),$$

and

$$\chi(S^2) = 2 - 1 + 1 = 2 = 2 - 2g(S^2).$$

One can calculate the fundamental group and the homology from this polygon. $\pi(M)$ is generated by the loops

$$a_1, a_2, \ldots, a_g, b_1, b_2, \ldots, b_g$$

with the relation

$$\prod_i a_i b_i a_i^{-1} b_i^{-1} = 1.$$

This loop is the boundary of the polygon and it can clearly be contracted on the manifold. A closer examination yields this to be the only relation.

To calculate H_1 we can use the fact that it is the abelianization of the fundamental group, hence

$$H_1(M) \cong \mathbb{Z}^{2g}.$$

We see $b_1 = 2g$ (b_1 is here the Betti number). In addition we see that the simplicial homology is independent on the triangulation chosen because the fundamental group does not depend on it.
In particular we obtain the identity

$$\chi(M) = b_0 - b_1 + b_2 = 2 - 2g.$$

We can also calculate H_1 directly by introducing a central point into the polygon, drawing lines to the vertices of the polygon and splitting each triangle into two (see Fig. 2.12 for the torus case).

Every closed 1-chain is homologically equivalent to a chain with components only on the boundary of the polygon. A closer examination shows that

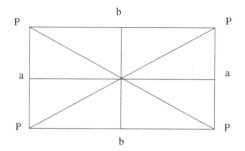

Fig. 2.12. Torus identification

it can be expressed in a_i and b_i and that they satisfy no nontrivial relations. Hence again we obtain the above result.

Because the normal form is fixed by the number of vertices of the polygon we see that the topological type of the Riemann surface we started with is given by the number g. Vice versa, by glueing such polygons we see there exists for every $g \geq 0$ a Riemann surface. This is the announced result of Sect. 1.3.

We have an *intersection product* in our homology. Let a_i and b_i be the generators of $H_1(M)$ given above (see Fig. 2.13).

If we represent them by loops we can define in a very visual way a intersection product

$$H_1(M) \times H_1(M) \rightarrow H_0(M) \cong \mathbb{Z}$$

induced by

$$a_i . a_j = 0, \quad a_i . b_j = -b_j . a_i = \delta_{ij}, \quad b_i . b_j = 0.$$

It is an alternating bilinear form. If

$$c = \sum_i n_i a_i + \sum_i m_i b_i, \quad c' = \sum_i n_i' a_i + \sum_i m_i' b_i$$

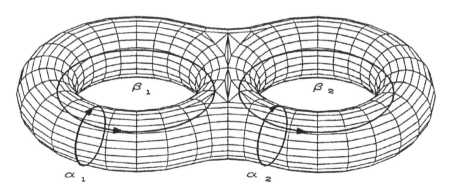

Fig. 2.13. Intersection product

are 2-cycles with $n_i, m_i, n'_i, m'_i \in \mathbb{Z}$ then their intersection is given by

$$c.c' = \begin{pmatrix} n & m \end{pmatrix} \cdot \begin{pmatrix} 0 & I \\ -I & 0 \end{pmatrix} \cdot \begin{pmatrix} n' \\ m' \end{pmatrix}$$

where I is the $g \times g$ identity matrix and $n, m, n', m' \in \mathbb{Z}^g$ are the coordinate vectors of c and c'. The $2g \times 2g$ matrix in the middle is called the intersection matrix. Of course it depends on the basis chosen for H_1. Bases in which it has this form are called *canonical homology bases*.

2.3 Universal Covering Space

In this section M and M' denote not necessarily compact Riemann surfaces. We study continuous maps

$$f: M \rightarrow M',$$

and we will see that if we make some assumptions on the structure of f, then there are not many of them.

Definition 2.11. *(a)* $f: M \rightarrow M'$ *is called a covering of M' if every point $x \in M'$ admits a neighbourhood U in such a way that*

$$f^{-1}(U) = \bigcup_j V_j, \quad V_j \cap V_k = \emptyset \quad \text{for } j \neq k$$

and

$$f_{|V_j} : V_j \cong U.$$

(b) f is called a ramified covering of M' if there is a discrete set of points A in M' such that the restriction

$$f_| : M \setminus f^{-1}(A) \rightarrow M' \setminus A$$

of f to $M \setminus f^{-1}(A)$ is a covering.

(c) A covering $f: M \rightarrow M'$ is called a universal covering of M' if M is simply connected.

By definition coverings are always surjective. The reason for calling a simply connected covering universal is given by the following:

Proposition 2.12. *Let $f: M \rightarrow M'$ be a universal covering. Let $g: M'' \rightarrow M'$ be another covering. If we choose a point $x_0 \in M'$ and a pair of points $y_0 \in f^{-1}(x_0)$, $z_0 \in g^{-1}(x_0)$ then there is exactly one map $k: M \rightarrow M''$ with*

$$f = g \circ k \quad \text{and} \quad z_0 = k(y_0).$$

The map k is a covering of M''.

From this it follows easily that up to canonical isomorphy the universal covering is unique.

Theorem 2.13. *Every compact Riemann surface admits a universal covering. In case of genus 0 (S^2) it is its own universal covering. In case of genus 1 its universal covering is the complex number plane $\mathbb{C}$. In case of higher genus it is the upper halfplane*

$$\mathcal{H} := \{\, z \in \mathbb{C} \mid \operatorname{Im} z > 0 \,\}.$$

Strictly speaking we have now left the field of topology because in this theorem we speak of Riemann surfaces with their complex structure. In fact these coverings are even analytic maps.

If we have a covering $f : M \to M'$ we can ask whether there are topological isomorphisms $\sigma : M \to M$ which cannot be seen on M' or, in exact terms, $f(\sigma(x)) = f(x)$ for all $x \in M$. We call these maps *covering transformations*.

Theorem 2.14. *Let M be the universal covering of M' then the group of covering transformations is isomorphic to the fundamental group of M'.*

The essential idea in the proof is the following fact. If p is a point on M' and q is a point on M lying above M' ($q \in f^{-1}(p)$) then a covering transformation takes q to a point q' also lying above p. Choose a path from q to q'. This path gives a loop in M' after applying f.

There are similar results connecting subgroups and factor groups of $\pi(M')$ with arbitrary coverings of M'.

Hints for Further Reading

In the book

[Fo]   Forster, O., *Riemannsche Flächen*, Springer, 1977.

or its English translation

[Foe]   Forster, O., *Lectures on Riemann Surfaces*, Springer, 1981.

one will find an extensive treatment of the topology of Riemann surfaces from the viewpoint of fundamental groups and covering spaces. The simplicial homology is more or less skipped. For this you can consult

[ST]   Seifert, H., Threlfall, W., *Lehrbuch der Topologie*, Chelsea, New York.

You find there the explicit reduction of every two-dimensional manifold to the standard polygon model. You will also find this in a more recent textbook

[St]   Stillwell, J., *Classical Topology and Combinatorial Group Theory*, Springer, 1980.

In fact for arbitrary spaces one uses today other homology theories. One example is the so-called *singular homology*. It is not so easy to visualize but much easier to handle in general constructions. Fortunately there is a theorem which says that in the cases which are of interest to us it coincides with the simplicial homology. You can find the singular homology in every modern textbook on algebraic topology, see for example

[Ma]   Massey, W.S., *Singular Homology Theory*, Springer, 1980.

There is another important generalization: *sheaf cohomology*. The definition of it is even more technical but one is also able to cover analytic or algebraic aspects with such methods. In the book of Forster you find one representative of these theories, the so-called Čech cohomology. It is also explained in Sect. 8 of these lectures. There are a lot of books on the subject. But most of them are rather voluminous and abstract. A standard reference is

[God]   Godement, R., *Theorie des faisceaux*, Hermann, Paris, 1964.

3

Analytic Structure

Let X be a Riemann surface. As we know X looks locally like the complex numbers $\mathbb{C}$ (via coordinate patches). Starting from a big enough atlas we can always get a coordinate patch (U_p, φ_p) around every point p by translation and restriction such that

$$\varphi_p : U_p \to \mathcal{E}, \quad \text{with } \varphi_p(p) = 0$$

where $\mathcal{E}$ is an open circle.

We call z_p (or later on just simply z) the complex coordinate in $\mathcal{E}$. By identification of

$$z_p \text{ with } \varphi_p^{-1}(z_p)$$

this is also a local coordinate on X around the point p. We call these kind of coordinates *standard coordinates*. Of course they are not unique. For the following we assume in most cases these standard coordinates without further mention.

Remark: In some older German books they are also called "Ortsuniformisierende".

3.1 Holomorphic and Meromorphic Functions

Definition 3.1. *Let X be a Riemann surface, Y an open subset of X and $f : Y \to \mathbb{C}$ a complex-valued function on Y.*
The function f is called holomorphic in the coordinate patch $(U \cap Y, \varphi_p)$ if

$$f \circ \varphi_p^{-1} : \quad \varphi_p(U \cap Y) \subset \mathbb{C} \to \mathbb{C}$$

is holomorphic.
A function $f : Y \to \mathbb{C}$ is called holomorphic if f is holomorphic in every coordinate patch.
By $\mathcal{O}(Y)$ we denote the set of all holomorphic functions on Y.

M. Schlichenmaier, Analytic Structure. In: M. Schlichenmaier, An Introduction to Riemann Surfaces, Algebraic Curves and Moduli Spaces, Theoretical and Mathematical Physics, 31–41 (2007)
DOI 10.1007/978-3-540-71175-9_4

Due to the fact that holomorphy is defined via the coordinates, it is clear that all local features of the usual holomorphic functions on $\mathbb{C}$ are valid for the holomorphic functions on Riemann surfaces.

Definition 3.2. *Let Y be an open subset of X.*
A function f is called a meromorphic function on Y, if there exists a non-empty open subset $Y' \subset Y$ with

(a) $f : Y' \to \mathbb{C}$ is a holomorphic function,

(b) $A = Y \setminus Y'$ is a set of isolated points (called the set of poles),

(c)
$$\lim_{x \to p} |f(x)| = \infty \quad for\ all \quad p \in A.$$

By $\mathcal{M}(Y)$ we denote the set of all meromorphic functions on Y.

It is easily seen that $\mathcal{M}(Y)$ is a field. We call $\mathcal{M}(X)$ the *field of meromorphic functions* on the Riemann surface X. It is a field extension of $\mathbb{C}$ of transcendence degree 1. Later on we will study some examples.

Definition 3.3. *Let X and X' be Riemann surfaces. A map $f : X \to X'$ is called holomorphic if f is holomorphic in every pair of coordinate patches, i.e. let $x \in X, f(x) \in X'$, φ_1 a coordinate around x, φ_2 a coordinate around $f(x)$, then we require*
$$\varphi_2 \circ f \circ \varphi_1^{-1} : \quad W \subset \mathbb{C} \to \mathbb{C}$$
to be holomorphic, where it is defined (see Fig. 3.1).
*   We call f an analytic isomorphism if f is bijective and f and f^{-1} are holomorphic (this is just a repetition of the definition in Sect. 1).*

One of the main problems in the theory is the following:

Classify all Riemann surfaces up to analytic isomorphy in a "natural way".

The set of isomorphy classes should carry some geometric structure which exhibits the appearance of Riemann surfaces. For example a small variation of the complex structure should correspond to a "nearby" point in the set. Later on we will try to make this more precise.

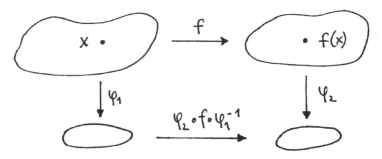

Fig. 3.1. Local form of a holomorphic map

Some remarks :

(1) Let f be a meromorphic function, p a point on X, (U, z) a standard coordinate patch around p. We can write locally:

$$f = \sum_{k=m}^{\infty} c_k z^k; \quad c_k \in \mathbb{C}, \quad c_m \neq 0.$$

The number m is called the *order* of f at the point p. It is easy to show that the order does not depend on the standard coordinate chosen for p.

If m is negative then p is a pole of f. We call $-m$ the *multiplicity* (or *order*) of the pole.

If $m \geq 0$ then f is holomorphic at p.

If $m > 0$ then f has a zero at p and we call m the *multiplicity* (or *order*) of the zero. Zeros of negative multiplicities are poles.

(2) Every holomorphic function, which is not identically zero, has only isolated points as zeros. If X is compact (which we assume) and f is a meromorphic function not identically zero, then f has only a finite set of poles and zeros. In fact, counted with multiplicities the number of zeros and the number of poles of a meromorphic function are the same.

(3) Let $F(z)$ be a complex polynomial. F regarded as a function $\mathbb{C} \to \mathbb{C}$ is holomorphic and either F is a constant or

$$\lim_{z \to \infty} |F(z)| = \infty.$$

With the usual inclusion $\mathbb{C} \subset \mathbb{P}^1$, we see $F(z) \in \mathcal{M}(\mathbb{P}^1)$.

(4) Let X be a Riemann surface, then every $f \in \mathcal{M}(X)$ defines, with the exception of the possibly finitely many poles of f, a map from

$$X \to \mathbb{C} \subset \mathbb{P}^1 = \mathbb{C} \cup \{\infty\}.$$

If we set for the poles p

$$f(p) = \infty,$$

we get an honest holomorphic map from

$$X \to \mathbb{P}^1.$$

Conversely every such map, which is not the map $f \equiv \infty$, defines a meromorphic function.

(5) Let f be a holomorphic map $X \to X'$. Written in local coordinates, how does f look like? Of course this depends on the coordinates chosen. But we can always choose coordinates z compatible with the complex structure around a and $b = f(a)$ such that for nonconstant f the representation is

$$f(z) = z^k, \qquad k \in \mathbb{N}.$$

k is unique, but the coordinates are not unique. In generalization of Remark 1, k is called the multiplicity of the value b with respect to f.

(6) Remark 5 has some important consequences for holomorphic maps. For example, nonconstant holomorphic maps are always open maps, i.e. the images of open sets are again open because the image of a full circle under the map z^k is again a full circle.

(7) Remark 6 implies

Theorem 3.4. *Let X, Y be Riemann surfaces, where X is compact, but Y is not necessarily assumed to be compact. If there is a map $f : X \to Y$ which is holomorphic and nonconstant, then Y is also compact and f is surjective.*

Proof. As f is open, $f(X)$ is an open subset of Y. On the other hand X is compact, so $f(X)$ is compact and hence closed. Because Y is connected the only nonempty closed and at the same time open set is the whole space, hence $f(X) = Y$, which shows the assertion. $\square$

(8) This theorem has some easy but important corollaries:
Let X be a Riemann surface (compact), f a globally holomorphic function, then f is a constant function. Or in other words,

$$\mathcal{O}(X) = \mathbb{C}.$$

For the proof, regard f as a holomorphic map $X \to \mathbb{C}$. Assume f to be nonconstant. Then by Theorem 3.4 the function f will be surjective and the image $\mathbb{C}$ has to be compact. This is a contradiction.

(9)
$$\mathcal{M}(\mathbb{P}^1) = \mathrm{Quot}(\mathbb{C}[z]).$$

In other words, every meromorphic function on the Riemann sphere is a rational function (i.e. the quotient of two polynomials).

Proof. Take $f \in \mathcal{M}(\mathbb{P}^1)$. But f has only finite many poles. Let n be the number of poles. Restricted to $\mathbb{C}$ we have a usual meromorphic function. We assume there is no pole at ∞. (Otherwise we take the function $f' = 1/f$.) If $a \in \mathbb{C}$ is a pole we have

$$f(z) = h(z) + g(z), \quad h(z) = \sum_{j=-m}^{-1} c_j(z-a)^j, \quad g \text{ is holomorphic at } a.$$

$h(z)$ is the principal part of $f(z)$ at a. Now

$$\lim_{z \to \infty} h(z) = 0,$$

so we can extend h to a meromorphic function on $\mathbb{P}^1$ by setting $h(\infty) = 0$. For every pole a_k we get a h_k. The function

$$f - h_1 - h_2 - \cdots - h_n \in \mathcal{M}(\mathbb{P}^1)$$

has no poles anymore, hence it is a holomorphic function. Due to the fact that $\mathbb{P}^1$ is compact, it is a constant, which shows the assertion. □

(10) A last remark on the classification problem:
Let $f : X \to Y$ be a holomorphic map and g a meromorphic function on Y, then the pullback

$$f^*(g) = g \circ f$$

is a meromorphic function on X. If $X \cong Y$ (analytically isomorphic), then this implies $\mathcal{M}(X) \cong \mathcal{M}(Y)$ (algebraic isomorphic as fields). The converse is also true, but this is a deeper result.

3.2 Divisors and the Theorem of Riemann–Roch

A divisor is nothing mysterious. A *divisor* is just an element of the free abelian group generated by the points of X. We denote this group by $\mathrm{Div}(X)$. Essentially we assign to every point an integer, where in all cases except finitely many this integer has to be zero. Let (in obvious abuse of notation)

$$D, D' \in \mathrm{Div}(X), \quad D = \sum_a n_a a, \quad D' = \sum_a m_a a.$$

(The sum is over all points $a \in X$.) Then we define

$$D + D' := \sum_a (n_a + m_a)a,$$

$$D \le D' \quad \text{if and only if} \quad n_a \le m_a \text{ for all } a.$$

For $f \in \mathcal{M}(X)$ a meromorphic function, a a point on X, recall that the *order* (or multiplicity) of f at a was defined as

$$\mathrm{ord}_a(f) := \begin{cases} 0, & f \text{ holomorphic at } a, f(a) \ne 0 \\ k, & f \text{ has a zero of multiplicity } k \text{ at } a \\ -k, & f \text{ has a pole of multiplicity } k \text{ at } a \\ \infty, & f \text{ is identically zero.} \end{cases}$$

To every meromorphic function $f \not\equiv 0$ we can assign a divisor

$$(f) = \sum_a \mathrm{ord}_a(f)a.$$

We call such divisors *principal divisors*. It is clear that

$$(f \cdot g) = (f) + (g), \qquad \left(\frac{1}{f}\right) = -(f).$$

Hence the principal divisors are a subgroup of $\mathrm{Div}(X)$. We say two divisors D, D' are *linearly equivalent* $D \sim D'$ if their difference is a principal divisor

$$D - D' = (f)$$

with an appropriate f. We use the name *divisor class* to denote the linear equivalence class of a divisor.

One of the main questions at this point is: Which divisors are principal divisors?

Or: Are there functions with certain prescribed poles and zeros. The theorem of Riemann–Roch will give us a first partial answer.

First we have to define the *degree* deg of a divisor

$$\deg : \mathrm{Div}(X) \ \to \ \mathbb{Z}; \qquad \sum_a n_a a \mapsto \sum_a n_a.$$

This is a homomorphism of groups. The set of divisors of degree 0 (denoted by $\mathrm{Div}^0(X)$) is a subgroup. This subgroup contains the subgroup of principal divisors, because every meromorphic function has the same number of zeros as poles.

The quotient of all divisors by the principal divisors is called *Picard group* $\mathrm{Pic}(X)$. The quotient $\mathrm{Div}^0(X)$ by the principal divisors is called the *restricted Picard group* $\mathrm{Pic}^0(X)$.

Let D be an arbitrary divisor, we define

$$L(D) := \{f \in \mathcal{M}(X) \mid (f) \geq -D\}.$$

If for example $n \cdot a$ with $n \geq 0$ is a term of D, then the $f \in L(D)$ are allowed to have a pole of at most multiplicity n at the point a. If $-n \cdot a$ is a term of D, then the $f \in L(D)$ has to have a zero of multiplicity at least n.

Obviously $L(D)$ is a vector space over the complex numbers. It is finite dimensional. We set

$$l(D) = \dim L(D).$$

In certain cases we are able to determine $l(D)$ directly.

If $\deg D < 0$ then $\deg -D > 0$. Hence $l(D) = 0$, because $\deg f = 0$ if $f \not\equiv 0$ (deg respects the ordering of the divisors).

We have $l(0) = 1$, because only constants have no poles.

The following theorem of Riemann–Roch gives us now a formula for $l(D)$:

Theorem 3.5. *(Riemann–Roch theorem)*

$$l(D) - l(K - D) = 1 - g + \deg D.$$

where g is the topological genus, K a canonical divisor (in Section 4 we will describe K with the means of differentials).

As an example of how to extract information from this formula let us calculate $l(K)$ and $\deg(K)$. We set $D = 0$ and get

$$1 - l(K) = 1 - g + \deg 0,$$

hence

$$l(K) = g.$$

We set $D = K$ and get

$$l(K) - l(0) = 1 - g + \deg K,$$

hence

$$\deg K = 2g - 2.$$

This implies: for

$$\deg D \geq 2g - 1$$

we have

$$\deg(K - D) < 0$$

and hence

$$l(K - D) = 0.$$

Finally we get the following results:

$$l(D) \begin{cases} = 0, & \text{if } \deg D < 0 \\ \geq 1 - g + \deg D, & \text{if } 0 \leq \deg D < 2g - 1 \\ = 1 - g + \deg D, & \text{if } \deg D \geq 2g - 1 \end{cases}$$

and the following existence theorem.

Theorem 3.6. *Let X be a compact Riemann surface of genus g, a a point on X. Then there exists a nonconstant function f on X, which has at a a pole of order at most $g + 1$ and is holomorphic elsewhere.* $\square$

Proof. Take

$$D = (g + 1)a, \quad \text{hence } \deg D = g + 1.$$

Riemann–Roch yields

$$l(D) \geq 1 - g + \deg D = 1 - g + g + 1 = 2.$$

Hence there are other functions in $L(D)$ besides the constant functions. As per definition of D one of these functions fulfils the assertion of the theorem.

$\square$

3.3 Meromorphic Functions on the Torus

Let ω_1, ω_2 be complex numbers which are linearly independent over the real numbers.

$$\Gamma := \{n\omega_1 + m\omega_2 \mid n, m \in \mathbb{Z}\}$$

is a two-dimensional lattice and

$$T = \mathbb{C}/\Gamma$$

is a torus with the induced complex structure coming from $\mathbb{C}$. Of course this structure depends very much on the lattice Γ.

If we consider another lattice

$$\Gamma' := \left\{n + m\frac{\omega_2}{\omega_1} \mid n, m \in \mathbb{Z}\right\}$$

and the associated torus T' we get a well-defined map $\Phi : T' \to T$, which is an analytic isomorphism

$$\bar{z} = z + \left(n + m\frac{\omega_2}{\omega_1}\right) \quad \mapsto \quad \omega_1 z + n\omega_1 + m\omega_2 = \overline{\omega_1 z} = \Phi(\bar{z}).$$

Essentially this is multiplication by ω_1. We see from the classification viewpoint that it is enough to consider only lattices of the type Γ'. Hence we assume for the following that Γ is already of this type.

In Γ we are able to choose the generator

$$\tau := \omega_2/\omega_1$$

such that its imaginary part is strictly positive. In this way we are able to restrict ourselves to the lattices of the form given in Fig. 3.2.

Let f be a meromorphic function on the complex plane. We call f doubly periodic (with respect to Γ) if

$$f(z + n + m\tau) = f(z), \quad \forall n, m \in \mathbb{Z}.$$

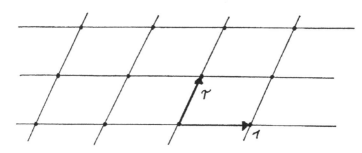

Fig. 3.2. A two-dimensional lattice

Such f defines in the obvious way an element $\bar{f} \in \mathcal{M}(T)$. Vice versa, given $g \in \mathcal{M}(T)$ we get by defining $f(z) := g(z + \Gamma)$ a function $f \in \mathcal{M}(\mathbb{C})$ which is doubly periodic and satisfies $\bar{f} = g$. Hence the doubly periodic meromorphic functions are the meromorphic functions on the torus. These functions are also called elliptic functions.

We now apply the usual residue theorem for such f along the boundary of the fundamental region (every point on the torus corresponds exactly to one point in the fundamental region)

$$\mathcal{F} := \{z \in \mathbb{C} \mid z = a + b\tau, \quad 0 \leq a, b < 1, a, b \in \mathbb{R}\},$$

$$\frac{1}{2\pi i} \int_{\partial \mathcal{F}} f(z)\, dz = \sum_{a \in \mathcal{F}} \mathrm{res}_a(f).$$

Due to the doubly periodicity of f we get

$$\int_{\partial \mathcal{F}} f(z)\, dz = 0,$$

hence also

$$\sum_{a \in \mathcal{F}} \mathrm{res}_a(f) = 0.$$

Finally we get the following fact:
There is no function on the torus which has exactly one pole of order 1. (Such a function would have a nonzero residuum.)

To describe all functions on the torus we use Weierstraß $\wp$-function

$$\wp(z) := \frac{1}{z^2} + {\sum_{w \in \Gamma}}' \left(\frac{1}{(z - w)^2} - \frac{1}{w^2} \right).$$

The $'$ denotes that we leave out 0 from the summation. This series converges for all $z \notin \Gamma$ to a holomorphic function. It has a pole of second order at the lattice points. It is doubly periodic.[1] If we differentiate we get

$$\wp'(z) = - \sum_{w \in \Gamma} \frac{2}{(z - w)^3}.$$

$\wp'$ is obviously doubly periodic and has poles of order 3 at the lattice points.

Of course $\wp$ and $\wp'$ are linearly independent, but they are not algebraically independent. Let 0 be the point on T corresponding to the lattice points in $\mathbb{C}$. We look at the divisors $n[0]$. Riemann–Roch tells us the dimension $l(n[0])$ of the vector space of meromorphic functions, which have a pole of order at most n at 0 and are holomorphic elsewhere:

---

[1] See Hurwitz/Courant, [HC], p. 161 for these results.

$$l(n[0]) = \begin{cases} 1, & n = 0 \\ n, & n \geq 1. \end{cases}$$

Hence if we list the known functions in $L(n[0])$ (1 is the constant function)

$$
\begin{aligned}
L(0[0]) &\supseteq <1> \\
L(1[0]) &\supseteq <1> \\
L(2[0]) &\supseteq <1, \wp> \\
L(3[0]) &\supseteq <1, \wp, \wp'> \\
L(4[0]) &\supseteq <1, \wp, \wp', \wp^2> \\
L(5[0]) &\supseteq <1, \wp, \wp', \wp^2, \wp\wp'> \\
L(6[0]) &\subseteq <1, \wp, \wp', \wp^2, \wp\wp', \wp^3, \wp'^2>,
\end{aligned}
\tag{3.1}
$$

we see there must be a nontrivial linear relation among the elements occurring in the last line. If we compare some of the first coefficients of the power series for $\wp$ and $\wp'$ we get the relation

$$(\wp')^2 = 4\wp^3 - g_2\wp - g_3, \tag{3.2}$$

with

$$g_2 = 60 {\sum_{w \in \Gamma}}' \frac{1}{w^4} = {\sum_{n,m \in \mathbb{Z}}}' \frac{1}{(n + m\tau)^4},$$

$$g_3 = 140 {\sum_{w \in \Gamma}}' \frac{1}{w^6} = {\sum_{n,m \in \mathbb{Z}}}' \frac{1}{(n + m\tau)^6}.$$

The g_2, g_3 are called *Eisenstein series*. If the lattice is fixed then they are constants. But they vary with the lattice. For our normalized lattices they are holomorphic functions in the variable τ.

In fact, we can substitute $\supseteq$ by $=$ in (3.1).

Starting with the Eisenstein series we are able to construct the *discriminant function*

$$\Delta(\tau) = g_2{}^3(\tau) - 27g_3{}^2(\tau).$$

As it is possible to show that $\Delta(\tau) \neq 0$ we define

$$j(\tau) = 1728 \cdot \frac{g_2{}^3(\tau)}{\Delta(\tau)}.$$

This j function will help us later on to classify the isomorphy classes of tori. It is called the *elliptic modular function*.

Theorem 3.7. *The field of meromorphic functions on the torus or equivalently the field of doubly–periodic (i.e. elliptic functions) can be given as*

$$\mathcal{M}(T) = \mathbb{C}(\wp, \wp')$$

with the relation (3.2) or in a more algebraic notation as

$$\mathcal{M}(T) \simeq \mathbb{C}(\mathbf{X})[\mathbf{Y}]/(\mathbf{Y}^2 - 4\mathbf{X}^3 + g_2\mathbf{X} + g_3).$$

Proof. Let $f \in \mathcal{M}(T)$ be given with a pole of order m at the point $\bar{a} \neq \bar{0}$.

$$g(\bar{z}) = f(\bar{z}) \cdot (\wp(\bar{z}) - \wp(\bar{a}))^m$$

is now another function which has this pole removed. By induction we reach a function having only poles at $\bar{0}$. Subtracting complex multiples of $\wp$ and $\wp'$ we get finally an everywhere holomorphic function on the torus, hence a constant. Working backwards we get the claim. $\square$

Hints for Further Reading

The book of Forster

[Fo]   Forster, O., *Riemannsche Flächen*, Springer, 1977.

or its English translation

[Foe]   Forster, O., *Lectures on Riemann Surfaces*, Springer, 1981.

covers the analytic aspect of the theory. Chap. 1 contains an extensive treatment of holomorphic functions, meromorphic function, differentials and integration. In Chap. 2 you also find the theorem of Riemann–Roch. Alas, he uses the cohomological method to formulate and prove it. Hence you have to learn Čech cohomology, which is developed in the book to understand the proof.

[FK]   Farkas, H.M., Kra, I., *Riemann Surfaces,* Springer, 1980.

is another very good book. The authors are rather sketchy on the basics of the analytic theory. But they prove Riemann–Roch without using cohomology. In the book you will find a lot of important topics which could not be covered in the lecture. Just to name some: uniformization, Greens functions, metrics on Riemann surfaces, automorphisms of Riemann surfaces and hyperelliptic Riemann surfaces. After the lectures you should be able to study these topics without too much difficulty.

On the specialities for one-dimensional tori a good reference is

[HC]   Hurwitz, A., Courant, R., *Allgemeine Funktionentheorie und elliptische Funktionen*, Springer, 1964.

4

Differentials and Integration

4.1 Tangent Space and Differentials

We have to recall in a condensed manner the notion of the tangent space of
a differentiable manifold M at a point $a \in M$. We use the simplest definition
(from the algebraic viewpoint).[1]

Definition 4.1. *Denote by $\mathcal{E}_a$ the algebra of germs of differentiable functions
at a (essentially this is the set of differentiable functions which are defined in
a neighbourhood of a, where the neighbourhood may depend on the function).*

(a) A point derivation D_a at a is a $\mathbb{R}$-linear map

$$\mathcal{E}_a \to \mathbb{R}$$

satisfying the product rule of differentiation

$$D_a(f \cdot g) = D_a(f) \cdot g(a) + f(a) \cdot D_a(g).$$

*(b) The tangent space $\mathbf{T}_a(M)$ of M at the point a is the vector space of all
point derivations at a with pointwise addition and multiplication with scalars.*

*Everything above extends in an obvious way to complex-valued functions and
derivations.*

It is possible to extend the notion of a derivation from the point to an open
set U or even the whole manifold M by allowing the point a to vary in the
set U (or M).

Let $(x_1, x_2, \ldots, x_n)$ be a set of local coordinates at a point a in U. Assume
furthermore that U is small enough such that this is a set of coordinates for
all points $b \in U$. Locally every function f can be expressed as a function of
x_i. The usual set of partial derivative operators

---

[1] Concerning the equivalence to more geometric definitions see Bröcker/Jänich,
[BJ], p. 17.

M. Schlichenmaier, Differentials and Integration. In: M. Schlichenmaier, An Introduction to
Riemann Surfaces, Algebraic Curves and Moduli Spaces, Theoretical and Mathematical Physics,
43–52 (2007)
DOI 10.1007/978-3-540-71175-9_5

$$\frac{\partial}{\partial x_i}, \quad i = 1, 2, \ldots, n$$

are derivations. In fact they form a basis of the tangent space at the point a. Even more, all local derivatives are generated by them as a module over the ring of arbitrary real functions. For the following we restrict ourselves to the derivations generated by $\mathcal{E}(U)$-functions

$$D_x = \sum_{i=1}^{n} a_i(x) \frac{\partial}{\partial x_i}, \quad a_i \in \mathcal{E}(U). \tag{4.1}$$

Recall that $\mathcal{E}(U)$ is the algebra of (infinitely often) differentiable functions on U. The functions a_i in (4.1) are differentiable functions. Such derivations are called local vector fields. A global vector field D is essentially a map from M to the disjoint union of all tangent spaces at every point:

$$D: \quad M \quad \rightarrow \quad \bigcup_{a \in M} \mathbf{T}_a(M),$$

where we require (besides some differentiability) that the point a is mapped to some element of $\mathbf{T}_a(M)$. Differentiability of the vector field D means that if we take a cover of M by coordinate neighbourhoods $\{U_\alpha\}_\alpha \in J$ then locally over U_α the map D can be expressed in the form (4.1) with $a_i \in \mathcal{E}(U_\alpha)$. The formula (4.2) will give the relation between the local representing function in two different coordinates. Local vector fields might be defined only for open subsets.

After choosing a basis, every local vector field can be represented by a n-vector of differentiable functions.

The elements of the dual vector space to the tangent space are the differential forms (1-forms, 1-differentials are other names). The differentials $dx_i, i = 1, 2, \ldots, n$ are defined as the dual basis with respect to $\partial/\partial x_i$:

$$dx_j\left(\frac{\partial}{\partial x_i}\right) = \delta_{ji}.$$

Locally, differential forms are given by

$$\omega = \sum_{i=1}^{n} \alpha_i(x) dx_i, \quad \alpha_i \in \mathcal{E}(U).$$

For $f \in \mathcal{E}(U)$ we can define a differential df by the condition

$$df(D) = D(f),$$

for all vector fields D on U, or in local coordinates x_i

$$df = \sum_{i=1}^{n} \frac{\partial f}{\partial x_i} dx_i.$$

As remarked earlier derivations and differentials can be represented locally in the coordinate patches by a n-vector of differentiable functions. Of course these functions depend on the coordinates chosen.

Let $y = \phi(x)$ be a change of coordinates, D a local derivation defined on both coordinate patches. Hence

$$D|_x = \sum_i a_i(x)\frac{\partial}{\partial x_i} = D|_y = \sum_j b_j(y)\frac{\partial}{\partial y_j}.$$

y_k is also a function which is defined at the intersection of both patches. We calculate

$$D(y_k) = \sum_i a_i(x)\frac{\partial y_k}{\partial x_i} = \sum_j b_j(y)\frac{\partial y_k}{\partial y_j} = \sum_j b_j \delta_{kj} = b_k(y).$$

Hence necessarily

$$b_k(y) = \sum_i a_i(x)\frac{\partial y_k}{\partial x_i}, \quad \text{with } y = \phi(x).$$

We set

$$A = \left(\frac{\partial y_k}{\partial x_i}\right)_{\substack{k=1,\dots,n \\ i=1,\dots,n}}$$

$$a(x) = {}^t(a_1(x), a_2(x), \dots, a_n(x)), \qquad b(y) = {}^t(b_1(y), b_2(y), \dots, b_n(y)),$$

and get

$$b(y) = {}^t A(x) \cdot a(x). \tag{4.2}$$

By linear algebra the representation with respect to the dual differentials transforms with

$$\beta(y) = A^{-1}(x) \cdot \alpha(x). \tag{4.3}$$

There is a rule to memorize the last relation. It is easy to show that dy_k as dual vector to $\partial/\partial y_k$ is the same as dy_k as differential of the function y_k, hence

$$dy_k = \sum_i \frac{\partial y_k}{\partial x_i} dx_i.$$

$\partial y_k/\partial x_i$ has to be compensated by the transformation of the functions representing the differentials.

There is a more general concept: In this concept vector fields are differentiable sections of the tangent bundle. They form the associated locally free sheaf. The matrices ${}^t A(x)$ are the defining transition matrices for this bundle.

In the same way differential forms are sections of the cotangent bundle. This bundle has transition matrices $A^{-1}(x)$. If you do not know this concept do not bother at the moment. We will define it in Chap. 8.

Up to now everything was over the real numbers. Our Riemann surfaces have a complex structure. So it is useful to consider complex functions, complex derivations, complex differentials and so on. Let X be a Riemann surface, (x, y) some local coordinates, $z = x + iy$ the local complex coordinate. The $\mathbb{C}$-derivations are generated by

$$\frac{\partial}{\partial x}, \quad \frac{\partial}{\partial y}.$$

On the other hand

$$\frac{\partial}{\partial z} = \frac{1}{2}\left(\frac{\partial}{\partial x} - i\frac{\partial}{\partial y}\right), \quad \frac{\partial}{\partial \bar{z}} = \frac{1}{2}\left(\frac{\partial}{\partial x} + i\frac{\partial}{\partial y}\right)$$

are also a basis. This basis fits our situation better. In the same way we switch from dx, dy to

$$dz = dx + idy, \quad d\bar{z} = dx - idy.$$

By $\mathcal{E}^1(U)$ we denote the set of differentials with differentiable coefficient functions. $\mathcal{E}^1(U)$ is a module over $\mathcal{E}(U)$. Locally we can write differential forms

$$\omega = f(z)\, dz + g(z)\, d\bar{z}, \quad f(z), g(z) \in \mathcal{E}(U).$$

If in this representation $g(z) \equiv 0$ we call ω a (1,0)-form. If $f(z) \equiv 0$ we call ω a (0,1)-form. (Be careful, we allow for (1,0)-forms $\partial f/\partial \bar{z} \neq 0$.) Of course this definition makes sense only if it does not depend on the holomorphic coordinates chosen. But this is easily verified.

We call a (1,0)-form ω a *holomorphic differential form* if

$$\omega = f(z)dz, \quad f(z) \in \mathcal{O}(U)$$

in every coordinate patch U where ω is defined. In the analytic theory it is also useful to consider *meromorphic differential forms*. A global meromorphic differential ω is a holomorphic differential on $X \setminus \{a_1, a_2, \ldots, a_m\}$. In addition we require the local representation of ω around a_i be given by a meromorphic function. Around every point a we get

$$\omega = f(z)dz, \quad f(z) = \sum_{i=m}^{\infty} c_i z^i, \quad c_m \neq 0.$$

We define

$$\mathrm{ord}_a(\omega) = m, \quad \mathrm{res}_a(\omega) = c_{-1}.$$

An easy calculation[2] shows these definitions are independent of the coordinates chosen. Remark: the residuum of a function on a Riemann surface is not well defined.

---

[2] See Forster, [Fo], p. 58.

Theorem 4.2. *(Residue Theorem) Let ω be a meromorphic differential on the Riemann surface X, then*

$$\sum_{a \in X} res_a(\omega) = 0.$$

It follows immediately that there exist no meromorphic differentials with exactly one pole of order 1.

Given a meromorphic differential $\omega \neq 0$ we can assign to it a divisor

$$\omega \mapsto (\omega) := \sum_{a \in X} ord_a(\omega)a.$$

We call (ω) a *canonical divisor*. Its linear equivalence class is called the *canonical divisor class K*. This is the class of divisors which is used in Theorem 3.5 (Riemann–Roch). Implicit in the definition of the canonical class is the claim that all canonical divisors are equivalent. But this is easy to see. Let $\beta \neq 0$ be another meromorphic differential. Locally they are given by

$$\omega = f(z)dz \quad = \quad f'(z')dz' \quad \text{with } f'(z') = f(z)\frac{\partial z}{\partial z'}$$

$$\beta = g(z)dz \quad = \quad g'(z')dz' \quad \text{with } g'(z') = g(z)\frac{\partial z}{\partial z'}.$$

Hence there exists locally meromorphic functions

$$h(z) = \frac{f(z)}{g(z)}$$

which glue together to form a global function because we have

$$\frac{f'(z')}{g'(z')} = \frac{f(z)}{g(z)}, \quad z' = z'(z)$$

on the overlap. Hence $h \cdot \beta = \omega$ and

$$(h) + (\beta) = (\omega) \quad \text{resp.} \quad (\beta) \sim (\omega).$$

Concerning the existence of such a ω see Theorem 4.6 in Sect. 4.3.

As a general fact linearly equivalent divisors D and D' have isomorphic $L(D)$ and $L(D')$. With the same technique we are able to show the following isomorphy. Let $\omega \neq 0$ be a meromorphic differential. The vector space

$$L((\omega)) := \{f \in \mathcal{M}(X) \mid (f) \geq -(\omega)\}$$

is isomorphic to

$$\Omega(X) := \{\text{global holomorphic differentials}\}.$$

The isomorphism is given by

$$L((\omega)) \to \Omega(X), \quad f \mapsto f \cdot \omega.$$

In fact $f \cdot \omega$ is a holomorphic differential because

$$(f) \ge -(\omega) \implies (f \cdot \omega) = (f) + (\omega) \ge -(\omega) + (\omega) = 0.$$

Example 4.3. On the torus $z + a$ are local variables, where z is the variable in $\mathbb{C}$ and $a \in \mathbb{C}$ is a displacement. dz is a globally defined differential because

$$\frac{\partial z'}{\partial z} = \frac{\partial(z + a)}{\partial z} = 1.$$

Hence
$$(dz) = 0 = (1) \quad \text{and} \quad \deg(dz) = 0 = 2g(T) - 2$$

in accordance with Riemann–Roch.

Example 4.4. On $\mathbb{P}^1$ we have two coordinate patches and they are given by

$$z \quad \text{and} \quad w = \frac{1}{z}.$$

dz is a well-defined holomorphic differential on $\mathbb{P}^1 \setminus \{\infty\}$. The transformation

$$z = \frac{1}{w} \quad \text{implies} \quad dz = -\frac{1}{w^2} dw.$$

Hence dz is a globally defined meromorphic differential with

$$(dz) = -2[\infty] \quad \text{and} \quad \deg(dz) = -2 = 2g(\mathbb{P}^1) - 2.$$

4.2 Differential Forms of Second Order

These are sections of the second exterior power of the cotangent bundle. A local basis is given by
$$dx \wedge dy \quad \text{or} \quad dz \wedge d\bar{z}.$$

They are related by

$$dz \wedge d\bar{z} = (dx + i dy) \wedge (dx - i dy) = -2i \, dx \wedge dy.$$

With respect to this basis a 2-form ω is locally represented by functions f

$$\omega = f(z) \, dz \wedge d\bar{z}.$$

The functions are related by the transformation rule

$$f(z) = f'(z') \left\| \frac{\partial z'}{\partial z} \right\|^2.$$

2-forms are necessarily (1,1)-forms, because (2,0) and (0,2) forms vanish like all other higher forms.

Let $f \in \mathcal{E}(U)$, then the differential df is defined by (see above)

$$df = \frac{\partial f}{\partial z} dz + \frac{\partial f}{\partial \bar{z}} d\bar{z} = \partial f + \bar{\partial} f$$

with

$$\partial f := \frac{\partial f}{\partial z} dz, \quad \bar{\partial} f := \frac{\partial f}{\partial \bar{z}} d\bar{z}.$$

Hence d splits in $d = \partial + \bar{\partial}$. This splitting is of course independent of the holomorphic coordinates chosen. d is also called the *exterior differentiation*. We can extend d to the higher differentials also. Let ω be a differential 1-form locally described by

$$\omega = f(z) dz + g(z) d\bar{z},$$

then we define

$$d\omega := df \wedge dz + dg \wedge d\bar{z} = \frac{\partial f}{\partial \bar{z}} d\bar{z} \wedge dz + \frac{\partial g}{\partial z} dz \wedge d\bar{z} = \left(\frac{\partial g}{\partial z} - \frac{\partial f}{\partial \bar{z}} \right) dz \wedge d\bar{z}$$

$$\partial \omega := \partial f \wedge dz + \partial g \wedge d\bar{z} = \frac{\partial g}{\partial z} dz \wedge d\bar{z}$$

$$\bar{\partial} \omega := \bar{\partial} f \wedge dz + \bar{\partial} g \wedge d\bar{z} = -\frac{\partial f}{\partial \bar{z}} dz \wedge d\bar{z}$$

A remark for the reader who is unfamiliar to the calculus of exterior powers: if $dv \wedge dw$ is a 2-form then we have the relation

$$dw \wedge dv = -dv \wedge dw.$$

Hence

$$dz \wedge dz = d\bar{z} \wedge d\bar{z} = 0$$
$$d\bar{z} \wedge dz = -dz \wedge d\bar{z}$$

in the above calculation.

We call a differential form $\omega \in \mathcal{E}^1(U)$ *closed* if $d\omega = 0$. We call ω *exact* if $\omega = df$ with $f \in \mathcal{E}(U)$. Holomorphic forms are automatically closed. One easily calculates

$$d^2 = \partial^2 = \bar{\partial}^2 = 0.$$

Hence exact forms are closed. We define the *de Rham cohomology group* as

$$H^1_{DR}(X) := \frac{\{ \text{ closed 1 - forms } \}}{\{ \text{ exact 1 - forms } \}}.$$

Of course it is also possible to define in the same way for higher dimensional manifolds higher de Rham cohomology groups $H^k_{DR}(X)$.

Let $f \in \mathcal{E}(U)$, then we calculate

$$\partial\bar{\partial}f = \partial\left(\frac{\partial f}{\partial\bar{z}}d\bar{z}\right) = \frac{\partial^2 f}{\partial z\partial\bar{z}}dz \wedge d\bar{z} = \frac{1}{2i}\left(\frac{\partial^2 f}{\partial x^2} + \frac{\partial^2 f}{\partial y^2}\right)dx \wedge dy$$
$$= \frac{1}{2i}\Delta(f)dx \wedge dy.$$

Hence we can call a function f *harmonic* if $\partial\bar{\partial}f = 0$ without coming in conflict with the definition by using the usual laplacian. As you see holomorphic and anti-holomorphic functions are harmonic.

4.3 Integration

Differentials are connected with integration. The result of integrating a k-form over a k-dimensional real compact submanifold is always a number. We are able to replace the submanifold by k-simplices or even by k-chains by linear continuation.

0-forms: Of course 0-forms are functions and 0-simplices are points. Integration of a function f over a point p is evaluation of f at p

$$\int_p f = f(p).$$

1-forms: Let

$$\gamma : I := [0,1] \to X$$

be a path and

$$\omega \in \mathcal{E}^1(X)$$

be a 1-form. Without loss of generality we can consider γ as a 1-chain. If γ is supported in one coordinate patch we are able to define $\int_\gamma \omega$ in the usual calculus manner. For example let (x,y) be coordinates,

$$\omega = f(x,y)\,dx + g(x,y)\,dy$$

be a differential, then

$$\int_\gamma \omega = \int_0^1 \left(f(x(t),y(t))\frac{dx}{dt} + g(x(t),y(t))\frac{dy}{dt}\right) dt.$$

This expression is independent of the choice of coordinates. To define the integral for an arbitrary path γ we split γ into parts, each of them lying completely inside one coordinate patch, integrate along the parts and sum up the results. An equivalent definition is by using a partition of unity.

2-forms: If ω is a 2-form, then we can define the integral over a dimension 2 compact submanifold in the usual calculus sense. By a partition of unity we get it for arbitrary 2-chains.

One of the most important facts in integration is

Theorem 4.5. *(Stokes' Theorem) Let ω be a k-form, C a $(k+1)$-chain then*

$$\int_C d\omega = \int_{\partial C} \omega.$$

By Stokes' theorem the integration pairing

$$\mathcal{E}^k(X) \times C_k(X) \to \mathbb{C}, \quad (\omega, C) \mapsto \int_C \omega$$

$(C_k(X)$ denotes as usual the group of k-chains) defines an induced pairing

$$H_{DR}^k(X) \times H_k(X, \mathbb{Z}) \to \mathbb{C}.$$

If we consider $H_k(X, \mathbb{C})$, the homology with complex coefficients, then the pairing is nondegenerated, hence

$$(H_k(X, \mathbb{C}))^* = H_{DR}^k(X).$$

Of course all the above is not restricted to Riemann surfaces. It is valid in higher dimensions.

Especially important in our situation is the following fact. Let γ_1 and γ_2 be paths from the point $p \in X$ to $q \in X$, and ω a closed 1-form. In general we have to expect

$$\int_{\gamma_1} \omega \neq \int_{\gamma_2} \omega.$$

Let $\alpha_k, \beta_k, k = 1, \ldots, g$ be a basis of $H_1(X, \mathbb{Z})$, where g is the genus. Because $\gamma_1\gamma_2^{-1}$ is a closed path we get

$$\gamma_1\gamma_2^{-1} \sim \sum_k n_k\alpha_k + \sum_k m_k\beta_k, \quad n_k, m_k \in \mathbb{Z}$$

modulo boundaries. Hence

$$\int_{\gamma_1} \omega - \int_{\gamma_2} \omega = \sum_k n_k \int_{\alpha_k} \omega + \sum_k m_k \int_{\beta_k} \omega.$$

In particular, paths which are equivalent in the homology yield the same result if we integrate along them.

Finally let us come to the question of existence of differentials with certain properties.

Theorem 4.6. *(a) Let* $p \in X$, z *a coordinate around* p, n *a natural number* ≥ 2 *then there exists a meromorphic differential* ω *with pole* $1/z^2$ *at* p *and holomorphic elsewhere. In other words there exists a* ω *with*

$$(\omega) = -2[p] + D$$

where D is a positive divisor not containing the point p.

(b) Let $p_1, p_2 \in X$, z_1, z_2 *be local coordinates at* p_1, *resp.* p_2, *then there exists a meromorphic differential* ω *holomorphic at* $X \setminus \{p_1, p_2\}$ *and poles* $1/z_1$ *at* p_1 *and* $-1/z_2$ *at* p_2.

(c) Let $\alpha_k, \beta_k, k = 1, \ldots, g$ *be a canonical homology basis, then there exists a unique set of holomorphic differentials* $\omega_k, k = 1, \ldots, g$ *with*

$$\int_{\alpha_j} \omega_i = \delta_{i,j}.$$

Moreover these ω_k *are a basis of the* $\mathbb{C}$-*vector space of holomorphic differentials.*

The key to this existence theorem is the fact that holomorphic objects are harmonic objects with respect to the real laplacian. Hence the real and imaginary parts alone are harmonic. First one constructs harmonic objects (potential theory, Dirichlet problem and so on) and then constructs with them the required meromorphic objects. Also, in part (c) the standard polygon cutting of the Riemann surface is used to get a simply connected domain.[3]

Hints for Further Reading

See also the Hints for Further Readings for Chaps. 1 and 3. In

[BJ] Bröcker, Th., Jänich, K., *Einführung in die Differentialtopologie*, Springer, 1973.

the equivalence of the different concepts of tangent space is discussed.

---

[3] For more information see Farkas/Kra, [FK], pp. 30–62.

5

Tori and Jacobians

5.1 Higher Dimensional Tori

First we have to define higher dimensional tori. Let

$$w_1, w_2, \ldots, w_{2n} \in \mathbb{C}^n$$

be linearly independent vectors over the real numbers.

$$L := \mathbb{Z}w_1 + \mathbb{Z}w_2 + \cdots + \mathbb{Z}w_{2n}$$

is a discrete subgroup of $\mathbb{C}^n$. We call L a *lattice*. The quotient $\mathbb{C}^n/L$ with the induced complex structure coming from $\mathbb{C}^n$ is an n-dimensional complex manifold. It is called an n-dimensional *torus*. In addition it comes with a group structure also induced from $\mathbb{C}^n$. With this group structure it is an analytic Lie group.

Our goal is to embed a certain class of tori into some projective space. For this purpose we have to add some additional structures to them. We say $T = \mathbb{C}^n/L$ carries a *polarization H* if

- H is a positive definite hermitian form on $\mathbb{C}^n$,
- $E = \operatorname{Im} H$ is a integer-valued antisymmetric form on L.

If $\det E = 1$ calculated on the basis of the lattice we call T a *principally polarized torus*. Let us remark here that not every torus admits a polarization.

We say two polarized tori (T_1, H_1) and (T_2, H_2) are isomorphic if there exists an analytic isomorphism

$$\phi : T_1 \to T_2 \quad \text{with} \quad \phi(0) = 0$$

such that the pullback of H_2 equals H_1:

$$\phi^*(H_2) = H_1.$$

M. Schlichenmaier, Tori and Jacobians. In: M. Schlichenmaier, An Introduction to Riemann Surfaces, Algebraic Curves and Moduli Spaces, Theoretical and Mathematical Physics, 53–60 (2007)
DOI 10.1007/978-3-540-71175-9_6

In defining the pullback of H_2 we use the fact that every analytic isomorphism

$$\phi : \mathbb{C}^n/L_1 \rightarrow \mathbb{C}^n/L_2, \quad \text{with} \quad \phi(0) = 0$$

can be given by a linear isomorphism

$$\phi_\# : \mathbb{C}^n \rightarrow \mathbb{C}^n \quad \text{with} \quad \phi_\#(L_1) = L_2 \tag{5.1}$$

and vice versa. The pullback is defined by

$$\phi_\#^*(H_2)(x,y) = H_2(\phi_\#(x), \phi_\#(y)).$$

From the presentation (5.1) we can also conclude that the problem of describing the different isomorphy types of tori (without polarization) can be reduced to the classification of different lattices in a fixed $\mathbb{C}^n$ up to a change of basis in $\mathbb{C}^n$. This is due to the fact that a linear isomorphism of $\mathbb{C}^n$ is equivalent to a change of basis in $\mathbb{C}^n$. (Obviously we can restrict ourselves to an isomorphism with $\phi(0) = 0$ because by translation (which is an analytic isomorphism) we can always reach this condition.)

For the following we restrict ourselves to principally polarized tori. Let

$$\Lambda = (\hat{w}_1, \hat{w}_2, \cdots, \hat{w}_{2n})$$

be the $n \times 2n$ matrix where the columns are just the coordinate vectors of the basis of the lattice. By a change of basis in $\mathbb{C}^n$ and a change of the basis in the lattice we can, in the case of principally polarized tori, always reach the following form:

$$\Lambda = (I, P),$$

where I denotes the $n \times n$ identity matrix, and P is a symmetric matrix with positive definite imaginary part, and E restricted to the lattice with respect to this basis is given by

$$J = \begin{pmatrix} 0 & I \\ -I & 0 \end{pmatrix}.$$

Such a basis is called a *symplectic basis*.

Let $(T_1 = \mathbb{C}^n/L_1, H_1)$ and $(T_2 = \mathbb{C}^n/L_2, H_2)$ be two principally polarized tori. If ϕ is an isomorphism of principally polarized tori then the map $\phi_\#$ transforms a symplectic basis of the lattice L_1 into a symplectic basis of the lattice L_2. Conversely, if ϕ is an isomorphism of tori such that $\phi_\#$ transforms a symplectic basis of L_1 into a symplectic basis of L_2, then $\phi_\#$ respects the antisymmetric forms E_1 and E_2 on the lattices. In other words, $\phi^*(H_2)$ equals H_1 when restricted to the lattice L_1. But an antisymmetric form E on a lattice fixes by $\mathbb{R}$-linear continuation an antisymmetric form on $\mathbb{R}^{2n} = \mathbb{C}^n$, also denoted by E. Such an antisymmetric form determines uniquely the associated hermitian form by

$$H(x,y) = E(\mathrm{i}x, y) + \mathrm{i}E(x,y).$$

Because $\phi_\#$ respects E_1 and E_2 it will respect the polarizations H_1 and H_2.

We had already reduced the classification of tori to the classification of lattices. We can assume that the lattice (and the relevant basis vectors) is given in such a way that the $n \times 2n$ matrix of the lattice basis has the form

$$\Lambda = (I, P).$$

The lattice is fixed by the matrix P. But this P is not unique. If we change the basis in the lattice, then Λ will be multiplied by a matrix

$$M \in \text{Sl}(2n, \mathbb{Z})$$

from the right. Now $\Lambda \cdot M$ is not in the above standard form. By a change of basis in $\mathbb{C}^n$, which means multiplication of $\Lambda \cdot M$ from the left with a matrix

$$N \in \text{Gl}(n, \mathbb{C})$$

we again normalize to

$$\Lambda' = N \cdot \Lambda \cdot M = (I, P').$$

Now P and P' describe equivalent lattices.

In the case of polarization two choices of basis in the lattice are equivalent only if the antisymmetric form E is invariant under such a transformation. Hence not the whole group $\text{Sl}(2n, \mathbb{Z})$ is allowed.

We define the *symplectic group*

$$\text{Sp}(2n, \mathbb{Z}) := \{\, M \in \text{Mat}(2n, \mathbb{Z}) \mid \quad {}^t M \cdot J \cdot M = J \,\}.$$

It is easily verified that it is a subgroup of $\text{Sl}(2n, \mathbb{Z})$ which is invariant under transposition. This group is the allowed group of such transformations of the basis (see the calculation at the end of this paragraph). Let

$$M = \begin{pmatrix} A & B \\ C & D \end{pmatrix} \in \text{Sp}(2n, \mathbb{Z}), \quad A, B, C, D \in \text{Mat}(n, \mathbb{Z})$$

Then (I, P) is transformed into (I, P') after renormalizing in $\mathbb{C}^n$ with

$$P' = (A \cdot P + B)(C \cdot P + D)^{-1}.$$

In fact, the isomorphy classes of principally polarized tori are given by the set of matrices P as above up to this operation of the group $\text{Sp}(2n, \mathbb{Z})$.

5.2 Jacobians

Let us go back to our Riemann surface of genus g. To exclude trivial cases we assume $g > 0$. Let

$$\alpha_1, \alpha_2, \ldots, \alpha_g, \beta_1, \beta_2, \ldots, \beta_g$$

be a canonical homology basis (see Chap. 2). Let

$$\omega_1, \omega_2, \ldots, \omega_g$$

be the associated unique basis of the holomorphic differentials. By definition we have

$$\int_{\alpha_j} \omega_i = \delta_{i,j}, \quad i,j = 1, \ldots, g.$$

We use the integration along the β_k to define

$$\pi_{ij} = \int_{\beta_j} \omega_i, \quad \Pi = (\pi_{ij})_{\substack{i=1,\ldots,g \\ j=1,\ldots,g}}.$$

This Π is called the *period matrix*.

Proposition 5.1. Π *is a complex-valued symmetric $g \times g$ matrix with positive definite imaginary part.*

To proof this, one uses the so-called *Riemann bilinear relations*:

$$\iint_X \eta \wedge \rho = \sum_{i=1}^g \left(\int_{\alpha_i} \eta \int_{\beta_i} \rho - \int_{\alpha_i} \rho \int_{\beta_i} \eta \right) \quad \text{for } \eta, \rho \text{ closed forms,}$$

$$0 = \sum_{i=1}^g \left(\int_{\alpha_i} \eta \int_{\beta_i} \rho - \int_{\alpha_i} \rho \int_{\beta_i} \eta \right) \quad \text{for } \eta, \rho \text{ holomorphic forms,}$$

$$\mathrm{Im} \left(\sum_{i=1}^g \overline{\int_{\alpha_i} \eta} \cdot \int_{\beta_i} \eta \right) > 0 \quad \text{for } \eta \text{ a holomorphic form.}$$

Essentially these relations are gained by cutting X along α_k, β_k into the standard polygon of Chap. 2.[1]

Now

$$L := \mathbb{Z}^g \oplus \Pi \cdot \mathbb{Z}^g$$

is a lattice in $\mathbb{C}^g$. It is generated by the standard column vectors and the columns of the matrix Π.

Definition 5.2. *The g-dimensional torus defined by*

$$\mathrm{Jac}(X) := \frac{\mathbb{C}^g}{L}$$

is called the Jacobian of the Riemann surface X.

There is the following more "natural" definition of $\mathrm{Jac}(X)$ which does not make any reference to some basis. Because $\Omega(X)$ is a g-dimensional vector space we can replace $\mathbb{C}^g$ by the space of linear forms on the space of global holomorphic differentials

---

[1] See Farkas/Kra, [FK], p. 56.

$$\text{Hom}\,(\Omega(X),\mathbb{C}) = (\Omega(X))^*.$$

Now every $\gamma \in H_1(X,\mathbb{Z})$ defines such a linear form by its periods

$$\omega \;\mapsto\; \int_\gamma \omega.$$

In this way $H_1(X,\mathbb{Z})$ is embedded into $(\Omega(X))^*$ and we are able to define

$$\text{Jac}\,(X) := \frac{\text{Hom}\,(\Omega(X),\mathbb{C})}{H_1(X,\mathbb{Z})}. \tag{5.2}$$

This more invariant description is useful in the context of moduli problems.

We see that the Jacobian comes with an alternating bilinear form. It is the intersection pairing on the homology. It can be described with respect to a canonical basis by the matrix J which appeared earlier:

$$J = \begin{pmatrix} 0 & I \\ -I & 0 \end{pmatrix} \in \text{Sl}\,(2g,\mathbb{Z}). \tag{5.3}$$

This intersection form coincides with the imaginary part of the hermitian form on $(\Omega(X))^*$ induced by duality from the hermitian form

$$< \omega_a, \omega_b >:= \frac{i}{2} \int_X \omega_a \wedge \overline{\omega_b}$$

on $\Omega(X)$. Hence the Jacobian comes as principally polarized torus.

In the next step we want to embed X into its own Jacobian. For the Jacobian we use the explicit description (the first definition). We choose an arbitrary point p_0 and define a map

$$\phi : X \to \text{Jac}\,(X), \quad p \mapsto \phi(p) = \left(\int_{p_0}^{p} \omega_1, \int_{p_0}^{p} \omega_2, \cdots, \int_{p_0}^{p} \omega_g \right) \quad \text{mod } L.$$

Of course the values of the integrals are not well defined; they depend on the path chosen from p_0 to p. Luckily after factoring out the lattice we get a well-defined map. Of course ϕ depends on the basepoint p_0 chosen. But a different basepoint will only result in a global translation of the image.

Proposition 5.3. *The map ϕ is an embedding.*

The map is called *Jacobi map.*

What are the advantages of Jac (X)?

(1) There is a very explicit way to embed Jac (X) into some projective space by the so-called theta functions (see Chap. 6). Via the Jacobi map we get an embedding of X into projective space and can pull back functions, differentials and so on.

(2) We have

Theorem 5.4. *(Torelli's theorem) Let X and X' be Riemann surfaces, then*

$$X \cong X' \quad \text{if and only if} \quad \mathrm{Jac}\,(X) \cong \mathrm{Jac}\,(X)'$$

with their principal polarizations introduced above.

In contrast to the moduli space for curves there exists an explicit description of the moduli space of polarized tori (see Chap. 7).

(3) To a Riemann surface X of genus $g > 0$ the assignment of $\mathrm{Jac}\,(X)$ is a natural way to assign to every curve an abelian group variety (or in the analytic language an analytic Lie group). In the case of genus 1, where X is already a group, we have $X \cong \mathrm{Jac}\,(X)$.

(4) The Jacobian also describes the restricted Picard group which was defined as the group of divisors of degree 0 modulo the principal divisors.

To see this, let D be a divisor of degree 0. We can split D with an appropriate n:

$$D = \sum_{i=1}^{n} p_i - \sum_{i=1}^{n} q_i.$$

We choose paths from q_i to p_i and define

$$\psi : \mathrm{Div}^0(X) \to \mathrm{Jac}\,(X),$$

$$D \mapsto \psi(D) = \sum_{i=1}^{n} \left(\int_{q_i}^{p_i} \omega_1, \int_{q_i}^{p_i} \omega_2, \ldots, \int_{q_i}^{p_i} \omega_g \right) \quad \text{mod } L.$$

The theorems of Abel and Jacobi[2] imply the kernel of ψ is exactly the subgroup of principal divisors and ψ is surjective. Hence the restricted Picard group of X is isomorphic to the Jacobian of X. It is sometimes called *Picard variety*.

In view of the Torelli correspondence we should also examine the following problem. If we fix X and a canonical homology basis we will get a unique associated basis of the differentials and hence a unique matrix Π. The problem is that the canonical basis is not uniquely determined. If we choose another such basis we will expect some change in the period matrix. What are the allowed changes of basis for the homology? Let M be the $2g \times 2g$ matrix of the coordinate transformation. We denote by (n, m) the old coordinates and by (n', m') the new coordinates, where $n, n', m, m' \in \mathbb{Z}^g$. Then M is defined by

$$\begin{pmatrix} n \\ m \end{pmatrix} = M \cdot \begin{pmatrix} n' \\ m' \end{pmatrix}.$$

This should be a change of basis over $\mathbb{Z}$. Hence our matrix M has to be an element of $\mathrm{Sl}\,(2g, \mathbb{Z})$. In addition we require that the new basis should have the same intersection form. Let J be the matrix (5.3).

---

[2] Farkas/Kra, [FK], p. 92; Forster, [Fo], p. 152.

$$\left(n_1 \ m_1\right) \cdot J \cdot \begin{pmatrix} n_2 \\ m_2 \end{pmatrix} = \left(n_1' \ m_1'\right) \cdot {}^t M \cdot J \cdot M \cdot \begin{pmatrix} n_2' \\ m_2' \end{pmatrix}$$

$$= \left(n_1' \ m_1'\right) \cdot J \cdot \begin{pmatrix} n_2' \\ m_2' \end{pmatrix}.$$

Hence

$${}^t M \cdot J \cdot M = J.$$

This is just the condition which defines the elements of the symplectic group $\mathrm{Sp}\,(2g, \mathbb{Z})$. Some calculation shows that for the new period matrix Π'

$$\Pi' = (A \cdot \Pi + B) \cdot (C \cdot \Pi + D)^{-1},$$

where A, B, C, D are $g \times g$ matrices over $\mathbb{Z}$ with

$$N := \begin{pmatrix} A & B \\ C & D \end{pmatrix} \in \mathrm{Sp}\,(2g, \mathbb{Z}).$$

One can calculate N explicitly from the coordinate change matrix M. Hence a different canonical homology basis yields another Jacobian which is still in the same isomorphy class of principally polarized tori.

Hints for Further Reading

In the book

[FK]   Farkas, H.M., Kra, I., *Riemann Surfaces,* Springer, 1980.

you will also find a chapter on Jacobians and theta functions. Of course there are also other books going more into details.

[Mum]   Mumford, D., *Tata Lectures on Theta I, II,* Birkhäuser, Boston, 1983 (1984).

These two volumes contain the important results quoted in the lecture and many more interesting topics. It requires only a little algebraic geometry. In the second volume the Jacobians of hyperelliptic Riemann surfaces are discussed. Another book by the same author

[MuC]   Mumford, D., *Curves and Their Jacobians,* The University of Michigan Press, Ann Arbor, 1976.

is a beautiful exposition of the relation announced in the title.
If you are looking for some special results you should check

[Fa]   Fay, J.D., *Theta Functions on Riemann Surface*, Lecture Notes in Mathematics 352, Springer, 1973.

He discusses the map $X \to \mathrm{Jac}\,(X)$ and constructs via theta functions objects on the Riemann surface X, for example the so-called prime form. There you also find references to the more classical literature.

6

Projective Varieties

6.1 Generalities

One of the techniques to study geometric objects is to embed them into some simple spaces and to gain in this way information about them. Of course we are able to embed every Riemann surface X into $\mathbb{R}^3$ as the handle model. However, this embedding does not respect the complex structure and is hence of no use in this context. At a first guess you might try to embed X into some $\mathbb{C}^n$. But all your efforts would be in vain, because except for the points there are no compact complex submanifolds in $\mathbb{C}^n$. The reason for this is quite simple. Assume X to be embedded in $\mathbb{C}^n$. Let z_i be the ith coordinate function in $\mathbb{C}^n$. It is a holomorphic function on the whole space and by the definition of complex embedding also a holomorphic function on X. But holomorphic functions on a compact Riemann surface are necessarily constants. In other words X has to be a point. So we have to look for the next simplest space. This is the projective space $\mathbb{P}^n$ we defined in Chap. 1

Starting with $\mathbb{P}^n$ it is also quite useful to use the algebraic geometric language. Let f be a polynomial in $n+1$ variables which is homogeneous of degree k. Homogeneous means that all terms in the polynomial (the so-called monomials) are of the same degree. Examples are

$$\alpha \mathbf{X}_0 - \beta \mathbf{X}_1, \alpha, \beta \in \mathbb{C}, \quad \mathbf{X}_0^2 + \mathbf{X}_1^2 + \mathbf{X}_2^2, \quad \mathbf{X}_1^2 \mathbf{X}_2 - 4\mathbf{X}_0^3 + \mathbf{X}_0 \mathbf{X}_2^2 + \mathbf{X}_2^3 .$$

If

$$z = (z_0 : z_1 : \ldots : z_n)$$

is a point of $\mathbb{P}^n$ we can plug its homogeneous coordinates into the polynomial f. For example $z \in \mathbb{P}^2$ and

$$f = \mathbf{X}_0^2 + \mathbf{X}_1^2 + \mathbf{X}_2^2$$

yields

$$f(z) = z_0^2 + z_1^2 + z_2^2.$$

M. Schlichenmaier, Projective Varieties. In: M. Schlichenmaier, An Introduction to Riemann Surfaces, Algebraic Curves and Moduli Spaces, Theoretical and Mathematical Physics, 61–70 (2007)
DOI 10.1007/978-3-540-71175-9_7

The value of the polynomial at z is not well defined, because

$$z' = (\lambda z_0 : \lambda z_1 : \lambda z_2)$$

is another representation of the same point z . We calculate

$$f(z') = \lambda^{\deg f} f(z).$$

Due to the homogeneity at least the alternative $f(z) = 0$ or $f(z) \neq 0$ is well defined. Hence we are able to define

$$V_f := \{z \in \mathbb{P}^n \mid f(z) = 0\}$$

the *variety* defined by f as the set of zeros of the polynomial f. We define now a topology $\mathcal{T}$ in $\mathbb{P}^n$ requiring that $\mathcal{T}$ is the coarsest topology in which the V_f are closed sets or equivalently where the complements $D_f = \mathbb{P}^n \setminus V_f$ are open sets. Hence the closed sets are given as finite unions and arbitrary intersections of such V_{f_i}. We call this topology $\mathcal{T}$ *Zariski topology*. Sets which are closed (or open) in this topology are closed (or open) in the usual real topology. It has a number of quite unusual properties. For example it is not separated; furthermore every nonempty Zariski open set U is automatically dense in $\mathbb{P}^n$. This is a very important fact. If we are able to show a condition is an open condition and is stated in algebraic terms we know the condition holds either nowhere or nearly everywhere.

As a simple example let us determine the closed sets of $\mathbb{P}^1$. If $f \equiv 0$ then $V_f = \mathbb{P}^1$. If $\deg f = 0$, then f is vanishing nowhere, hence $V_f = \emptyset$. Now let

$$f = \sum_{i=0}^{k} a_i \mathbf{X}_0^i \mathbf{X}_1^{k-i}$$

be a homogeneous polynomial of degree $k > 0$. Assume first $(1,0)$ is not a zero of f. If we plug in the points $(z : 1)$ we get the expression

$$f(z) = \sum_{i=0}^{k} a_i z^i$$

which is zero exactly at k values for z counted with multiplicities . If $(1,0)$ is a zero we can divide out $\mathbf{X}_1$ a certain number of times to get a polynomial f' which has no zero at $(1,0)$. Hence we are reduced to the first case. Because only finite unions are allowed in constructing closed sets we get

Proposition 6.1. *Closed sets in $\mathbb{P}^1$ are exactly the sets consisting of finitely many points, the empty set and $\mathbb{P}^1$ itself.*

It is clear that every finite set of points is the zero set of a product of linear polynomials and hence closed.

Definition 6.2. *(a) A subset W of $\mathbb{P}^n$ is called a (projective) variety if it is the common set of zeros of finitely many homogeneous polynomials (not necessarily of the same degree).*
(b) A variety is called a linear variety if it is the zero set of linear polynomials.
(c) A variety W is called irreducible if every decomposition

$$W = V_1 \cup V_2$$

with V_1 and V_2 varieties implies

$$V_1 \subset V_2 \quad or \quad V_2 \subset V_1.$$

(d) A Zariski open set of a projective variety is called quasi-projective or sometimes just algebraic.

The following are a few remarks on the definitions. In (a) we could also have used arbitrarily many polynomials with the same effect. This is due to the result that the common zero set of arbitrarily many polynomials is always a zero set of finitely many polynomials.[1] Of course in definition (d) we define the Zariski topology on a projective variety W as the topology obtained by restricting the topology of $\mathbb{P}^n$ to W.

Sometimes it is useful to reserve the name "varieties" for irreducible varieties and simply use the name "algebraic sets" for the not necessarily irreducible ones, see Chap. 11.

Let us consider some examples. A line $\mathcal{L}_1$ in $\mathbb{P}^2$ is given as the zero set of a nonvanishing linear form

$$l_1 = a\mathbf{X}_0 + b\mathbf{X}_1 + c\mathbf{X}_2, \quad a, b, c \in \mathbb{C}.$$

Let $\mathcal{L}_2$ be another line given by the linear form

$$l_2 = a'\mathbf{X}_0 + b'\mathbf{X}_1 + c'\mathbf{X}_2, \quad a', b', c' \in \mathbb{C}.$$

The vector (a, b, c) is a multiple of (a', b', c') if and only if $\mathcal{L}_1 = \mathcal{L}_2$. Assume for the following that they are different. Clearly $\mathcal{L}_1$ and $\mathcal{L}_2$ are irreducible subvarieties. Their union $W = \mathcal{L}_1 \cup \mathcal{L}_2$ is again a subvariety. It is the union of the two lines, reducible and given as the zero set of the reducible polynomial

$$l_1 \cdot l_2.$$

The intersection $V = \mathcal{L}_1 \cap \mathcal{L}_2$ is the variety

$$\{z \in \mathbb{P}^2 \mid l_1(z) = 0, l_2(z) = 0\}.$$

---

[1] Or in more algebraic terms, the polynomial ring is a noetherian ring, which says that every ideal is generated by a finite number of elements.

Of course this is exactly the intersection point of the two lines. It cannot be described by just one polynomial.

If V is an irreducible variety we can define its *dimension* by the length n of a maximal chain of strict subvarieties which are irreducible:

$$\emptyset \subsetneq V_0 \subsetneq V_1 \subsetneq V_2 \subsetneq V_{n-1} \subsetneq V_n = V.$$

A variety of dimension 1 is called a curve, a variety of dimension 2 is called a surface and so on.

The *rational functions* on $\mathbb{P}^n$ are defined as quotients of two homogeneous polynomials in $n+1$ variables of the same degree:

$$h(\mathbf{X}) = \frac{f(\mathbf{X})}{g(\mathbf{X})}.$$

For every point z with $g(z) \neq 0$ we get a well-defined complex value and hence a function on $D_g = \mathbb{P}^n \setminus V_g$ in the usual sense. The rational functions are the equivalent of the meromorphic functions. In the affine coordinate patch U_0 (defined in Sect. 1.2) we can set $z_0 = 1$ and get a rational function as a quotient of polynomials in n variables without the homogeneity restriction. We calculated $\mathcal{M}(\mathbb{P}^1)$ in Chap. 3 and now we also see that they are the same as the rational functions on $\mathbb{P}^1$. This is not by coincidence, see Chow's theorem below.

The rational functions on a quasi-projective variety V are defined as quotients of homogeneous polynomials of the same degree:

$$h(\mathbf{X}) = \frac{f(\mathbf{X})}{g(\mathbf{X})} \quad \text{with} \quad g|_V \not\equiv 0.$$

(*Weil-*)*divisors* on a projective variety are again elements of the free abelian group generated by the 1-codimensional irreducible subvarieties.

An *algebraic map* between projective varieties X and X' is a map

$$f : X \to X',$$

where f can be given locally as a set of rational functions without poles.

Contrary to the case of manifolds, there is no need to exclude singularities from the beginning. A large amount of the theory works without requiring nonsingularity or only requiring some weaker conditions like normality. This is important for the study of the boundary of certain moduli spaces. But to see the connection with our Riemann surfaces we should define nonsingularity now.

Definition 6.3. *Let V be a projective variety of dimension r in $\mathbb{P}^n$ defined by the polynomials $f_1, f_2, \ldots, f_m$. Consider the $m \times (n+1)$ matrix*

$$M(\mathbf{X}) := \begin{pmatrix} \dfrac{\partial f_1}{\partial \mathbf{X}_0} & \dfrac{\partial f_1}{\partial \mathbf{X}_1} & \cdots & \dfrac{\partial f_1}{\partial \mathbf{X}_n} \\ \dfrac{\partial f_2}{\partial \mathbf{X}_0} & \dfrac{\partial f_2}{\partial \mathbf{X}_1} & \cdots & \dfrac{\partial f_2}{\partial \mathbf{X}_n} \\ \vdots & \vdots & \ddots & \vdots \\ \dfrac{\partial f_m}{\partial \mathbf{X}_0} & \dfrac{\partial f_m}{\partial \mathbf{X}_1} & \cdots & \dfrac{\partial f_m}{\partial \mathbf{X}_n} \end{pmatrix}.$$

We call a point z on V with

$$rank\ M(z) < n - r$$

a singular point. If V has no singular points it is called a nonsingular variety (sometimes also smooth variety).

This definition is somehow unsatisfactory because it is not defined as some intrinsic property of the variety. In fact there are some equivalent definitions via local rings, differentials and so on which are more intrinsic. With them it is easy to see that the nonsingularity does not depend on the embedding.

Nonsingular varieties (they are always irreducible) are compact complex manifolds. The converse is also true in the following sense.

Theorem 6.4. *(Theorem of Chow)*[2] *Let X be a compact complex manifold holomorphically embedded in $\mathbb{P}^n$. Then*

(a) *X is a nonsingular projective variety.*

(b) *Every meromorphic function on X is a rational function.*

(c) *Every meromorphic differential is a rational differential.*

(d) *Every holomorphic map between two varieties is algebraic.*

The Kodaira embedding theorem says every Riemann surface admits an embedding into some projective space. By Chow's theorem we get

Proposition 6.5. *Every Riemann surface is analytically isomorphic to a projective nonsingular curve (do not forget: our Riemann surfaces are always assumed to be compact).*

Chow's theorem says even more: we have a complete equivalence of the category of Riemann surfaces and the category of projective nonsingular curves, including maps, divisors, Riemann–Roch and so on.

6.2 Embedding of One-Dimensional Tori

Let $T = \mathbb{C}/L$ be a torus and

$$(\wp')^2 = 4\wp^3 - g_2\wp - g_3, \quad g_2{}^3 - 27g_3{}^2 \neq 0$$

---

[2] Griffiths/Harris, [GH], p. 166.

be the relation between the $\wp$ and $\wp'$ function, see Sect. 3.3. We define the following map:

$$\Psi : T \to \mathbb{P}^2, \qquad \bar{z} \mapsto \begin{cases} (\wp(z), \wp'(z), 1), & \bar{z} \neq \bar{0} \\ (0, 1, 0), & \bar{z} = \bar{0} \end{cases},$$

where $\bar{z} = z + L$. Recall that $\wp(z + L) = \wp(z)$.

In $\mathbb{P}^2$ we define the cubic curve V_f as the zero set of the following polynomial:

$$f(\mathbf{X}, \mathbf{Y}, \mathbf{Z}) = \mathbf{Y}^2 \mathbf{Z} - 4\mathbf{X}^3 + g_2 \mathbf{X} \mathbf{Z}^2 + g_3 \mathbf{Z}^3.$$

This cubic curve is also called an *elliptic curve*. We calculate

$$\left(\frac{\partial f}{\partial \mathbf{X}}, \frac{\partial f}{\partial \mathbf{Y}}, \frac{\partial f}{\partial \mathbf{Z}} \right) = (-12\mathbf{X}^2 + g_2 \mathbf{Z}^2 \ , \ 2\mathbf{Y}\mathbf{Z} \ , \ \mathbf{Y}^2 + 2g_2 \mathbf{X} \mathbf{Z} + 3g_3 \mathbf{Z}^2).$$

For a singular point $p = (x : y : z)$ all three entries have to vanish.
Case 1: $z = 0$. Hence

$$(-12x^2, 0, y^2) = (0, 0, 0)$$

which implies $y = 0$ and $x = 0$. But this is not a point of $\mathbb{P}^2$.
Case 2: $z \neq 0$. We can fix $z = 1$ and get $y = 0$ and

$$-12x^2 + g_2 = 0, \qquad 2g_2 x + 3g_3 = 0.$$

By multiplication we get

$$g_2{}^3 = 12x^2 g_2{}^2$$
$$9g_3{}^2 = 4x^2 g_2{}^2.$$

If there were a solution we would have necessarily

$$g_2{}^3 - 27g_3{}^2 = 0$$

in contradiction to our starting relation. Hence this cubic curve is nonsingular.

The differential equation for the $\wp$ function implies that the image of Ψ is a subset of the curve V_f. $\wp$ is a function on the torus, hence it takes every value of $\mathbb{P}^1$ equally often (calculated with multiplicity). It has a pole of order 2 at $0 \in T$ and nowhere else. Hence every value occurs two times. But $\wp$ is an even function, hence the two points are exactly $\wp(z)$ and $\wp(-z)$. Of course it could happen that $\bar{z} = -\bar{z} \in T$. But $\wp'$ is an odd function and hence will vanish for such z. In other words the multiplicity there is 2. If $\bar{z} \neq -\bar{z}$ then $\wp'$ separates the two points. This implies Ψ is injective. On the other hand for

$$(\eta, \rho, 1) \in V_f, \qquad \rho^2 = 4\eta^3 - g_2\eta - g_3$$

we can always find a z with

$$\wp(z) = \eta \quad \text{and} \quad \wp'(z) = \rho.$$

Hence Ψ is bijective. Because it is described by analytic functions it is an analytic isomorphism.

In Sect. 3.3 we showed

$$\mathcal{M}(T) = \mathbb{C}(\wp, \wp').$$

Via Ψ the function $\wp$ corresponds to the rational function $\mathbf{X}/\mathbf{Z}$ and $\wp'$ corresponds to $\mathbf{Y}/\mathbf{Z}$.

6.3 Theta Functions

We define *Siegel's upper half-space* as

$$\mathcal{H}_g := \{\Pi \in \text{Mat}\,(g, \mathbb{C}) \mid \quad \Pi = {}^t\Pi, \text{Im}\,\Pi > 0\}. \tag{6.1}$$

Of fundamental importance is the following:

Definition 6.6. *We call*

$$\vartheta(z, \Pi) = \sum_{n \in \mathbb{Z}^g} \exp(\pi\mathrm{i}\,{}^t n \Pi n + 2\pi\mathrm{i}\,{}^t n\,z)$$

with

$$z \in \mathbb{C}^g, \quad \Pi \in \mathcal{H}_g$$

the theta function.

Proposition 6.7. ϑ *is a well-defined holomorphic function on*

$$\mathbb{C}^g \times \mathcal{H}_g\,.$$

This is proved by a suitable dominating convergent series.[3] Of course we are not interested so much in functions on $\mathbb{C}^g$. Let us fix a $\Pi \in \mathcal{H}_g$ and see how to relate $\vartheta(z, \Pi)$ to the torus with principal polarization

$$T = \frac{\mathbb{C}^g}{L}, \quad L = \mathbb{Z}^g \oplus \Pi \cdot \mathbb{Z}^g \quad.$$

We cannot expect it to be periodic because it would then be a holomorphic function on T, and on T the only holomorphic functions are the constants as in the one-dimensional case. Instead it has a well-defined quasi-periodic behaviour under translation with lattice vectors.

---

[3] For this and the proofs of the following propositions see Mumford, [Mum], Theta I, p. 118.

Proposition 6.8.

$$\vartheta(z + m, \Pi) = \vartheta(z, \Pi), \qquad m \in \mathbb{Z}^g$$
$$\vartheta(z + \Pi m, \Pi) = \exp(-\pi i\, {}^t m \Pi m - 2\pi i\, {}^t m\, z) \cdot \vartheta(z, \Pi), \qquad m \in \mathbb{Z}^g.$$

In fact ϑ is the only function up to constant multiples with this quasi-periodicity. This follows from Fourier expansion.

Our aim is to embed the principally polarized tori into projective space via theta functions. So far we have defined only one theta function. To get more we have to expand the definition slightly. We choose a natural number l and look for holomorphic functions f with

$$f(z + m) = f(z), \qquad m \in \mathbb{Z}^g \tag{6.2}$$
$$f(z + \Pi m) = \exp(-\pi i l\, {}^t m \Pi m - 2\pi i l\, {}^t m\, z) \cdot f(z), \qquad m \in \mathbb{Z}^g.$$

We call f *quasi-periodic of weight l*.

First let us show that these are useful objects for our problem. Let $f_1, f_2, \dots, f_n$ be such functions. If there exists for every $z \in \mathbb{C}^g$ at least one f_i with $f_i(z) \neq 0$ then

$$\psi : \mathbb{C}^g \to \mathbb{P}^n, \qquad z \mapsto (f_0(z) : f_1(z) : \dots : f_n(z))$$

defines a holomorphic map. But this map satisfies

$$\psi(z + L) = \psi(z)$$

because translations by lattice vectors just change the coordinates of the image point by the same overall multiple and hence yield the same point in projective space. Therefore

$$\psi : \quad T \quad \to \quad \mathbb{P}^n$$

is a holomorphic map. We have only to choose l in such a way that there are enough functions with this quasi-periodicity that the above condition can be fulfiled. To get an embedding we also require that ψ is injective (it separates points) and the image is nonsingular (it separates tangent vectors). For this reason we define theta functions with characteristics $a, b \in \mathbb{Q}^g$:

$$\vartheta \begin{bmatrix} a \\ b \end{bmatrix}(z, \Pi) := \exp(\pi i\, {}^t a \Pi a + 2\pi i\, {}^t a(z + b)) \cdot \vartheta(z + \Pi a + b, \Pi).$$

We have

$$\vartheta = \vartheta \begin{bmatrix} 0 \\ 0 \end{bmatrix}.$$

A possible basis of the vector space of functions obeying the condition (6.2) is given in the case of $l = k^2$ by the set

$$h_{a,b}(z) := \vartheta \begin{bmatrix} a/k \\ b/k \end{bmatrix}(l\, z, \Pi), \qquad a, b \in \mathbb{Z}^g, z \in \mathbb{C}^g$$

with

$$a = (a_1, a_2, \ldots, a_g), \qquad b = (b_1, b_2, \ldots, b_g), \qquad 0 \leq a_i, b_i < k.$$

If we choose l high enough we get an embedding.

By this embedding, the theta functions correspond to the linear polynomials in $\mathbb{P}^n$ and the rational functions are given as quotients of polynomials of theta functions where the numerator and the denominator are of the same degree.

By now we know principally polarized tori are varieties. A closer examination yields

Theorem 6.9. *A torus is embeddable into projective space if and only if it admits a polarization. The embedding depends on the polarization chosen.*

With this embedding the analytic group law becomes an algebraic group law. Hence we speak of these embedded tori in the algebraic language as *abelian varieties*.

Hints for Further Reading

It is rather hard to make a useful suggestion because most of the books in algebraic geometry start from scratch by using the framework of commutative algebra. This is of course the right way to do algebraic geometry. A book which mixes the analytic and algebraic techniques and hence might be easier to read for somebody already acquainted with the analytic language is

[GH] Griffiths, Ph., Harris, J., *Principles of Algebraic Geometry*, Wiley, 1978.

In this book you also find an introduction to complex manifolds and other fundamental material like Kähler manifolds, Hodge theory and so on. If you are interested only in some small portion of it, it is not necessary to start from the beginning. Just enter the chapter you are interested in and check the special preliminaries you need as you go on.

In this 2nd edition, in the Chaps. 11 and 12 an introduction to modern algebraic geometry is given. Another book describing the algebraic geometry from the purely algebraic viewpoint is

[Ha] Hartshorne, R., *Algebraic Geometry*, Springer, 1977.

After a first chapter of about 60 pages where he uses the language of varieties we also used in our lectures, he skips to the modern language of schemes more often used nowadays.

More introductory texts in algebraic geometry are

[Ku] Kunz, E., *Einführung in die kommutative Algebra und algebraische Geometrie*, Vieweg, Braunschweig, 1980

or the English translation

[Kue] Kunz, E., *Introduction to Commutative Algebra and Algebraic Geometry*, Birkhäuser, Boston, 1985,

and

[Ke] Kendig, K., *Elementary Algebraic Geometry*, Springer, 1977.

There is a rather elementary book on curves which is also very useful for an introduction to algebraic geometry:

[Fu] Fulton, W., *Algebraic Curves*, Benjamin/Cummings, Reading, MA, 1969.

A more advanced book on curves is

[Ar] Arbarello, E., Cornalba, M., Griffiths, Ph., Harris, J., *Geometry of Algebraic Curves,* Vol. I, Springer, 1985.

The embedding of tori can be found in

[Mum] Mumford, D., *Tata Lectures on Theta I, II*, Birkhäuser, Boston, 1983 (1984).

7

Moduli Spaces of Curves

7.1 The Definition

In this section we use the word curve to denote a nonsingular projective curve. We also switch freely between the algebraic and the analytic language. We define the moduli space $\mathcal{M}_g$ as

$$\mathcal{M}_g := \{\text{ analytic isomorphy classes of Riemann surfaces of genus } g \}$$

or equivalently

$$\mathcal{M}_g := \{\text{ algebraic isomorphy classes of curves of algebraic genus } g \}.$$

In the algebraic setting we want to make only minimal reference to the topology, hence we introduce the *algebraic genus* as the dimension of the vector space of global algebraic differentials without poles. To simplify the notation we also call them holomorphic differentials. Of course over the complex numbers $\mathbb{C}$ the algebraic genus equals the topologically defined genus.

Up to now $\mathcal{M}_g$ is only a set and we should give it a geometric structure to make it into a space. The structure should reflect all possibilities in which the Riemann surfaces could show up. For example a small deformation of the complex structure should correspond to a small path in $\mathcal{M}_g$ and so on. Unfortunately to make this more precise we have to introduce some more notation.

Definition 7.1. *A family of curves (Riemann surfaces) of genus g over B is a surjective map*

$$\pi : \quad X \quad \to \quad B$$

where X, B are algebraic varieties (complex manifolds), π is an algebraic (analytic) map and the fibre

$$X_b = \pi^{-1}(b)$$

M. Schlichenmaier, Moduli Spaces of Curves. In: M. Schlichenmaier, An Introduction to Riemann Surfaces, Algebraic Curves and Moduli Spaces, Theoretical and Mathematical Physics, 71–86 (2007)
DOI 10.1007/978-3-540-71175-9_8

*is a curve (a Riemann surface) of genus g for every point $b \in B$.
In the case of genus 1 we require in addition that the map*

$$\sigma : \quad B \quad \rightarrow \quad X$$

*defined by assigning to every b the element 0 in X_b is an algebraic (analytic)
map. B is called base variety (manifold).*
*We call two families X, X' isomorphic, if they are families over the same base
variety B*

$$\pi : X \rightarrow B, \qquad \pi' : X' \rightarrow B$$

and if there exists a algebraic (analytic) isomorphism

$$\sigma : X \rightarrow X'$$

*such that $\pi' \circ \sigma = \pi$. In particular, this says that σ has to respect fibres and
that fibres over the same basepoint $b \in B$ are isomorphic.*

Let us first give an example of such a family for $g = 1$ in the analytic language.
Let U be an open subset of $\mathbb{C}$, $\mathcal{H}$ the upper half-space

$$\mathcal{H} := \{ z \in \mathbb{C} \mid \operatorname{Im} z > 0 \}$$

and

$$\tau : U \rightarrow \mathcal{H}$$

be an analytic function. Now we construct a family over U where the fibre
over the point z is exactly the torus defined by the lattice

$$\mathbb{Z} \oplus \tau(z) \mathbb{Z}.$$

Therefore we set

$$X := U \times \mathbb{C}/ \sim$$

where $\sim$ is defined as

$$(z, w) \sim (z', w') \quad \text{if and only if} \quad z' = z \text{ and } w' = w + n + \tau(z)m, \quad n, m \in \mathbb{Z}.$$

As π we take the projection on the first factor. It is not hard to see that the
conditions for being a family of Riemann surfaces are fulfiled.

If we have a family $\pi : X \rightarrow B$ (for arbitrary genus g) we can define a
map

$$\psi_{B,X} : \quad B \quad \rightarrow \quad \mathcal{M}_g$$

by assigning to $b \in B$ the isomorphy class of the fibre X_b over b. The geometric
structure on $\mathcal{M}_g$ should be such that $\psi_{B,X}$ is an algebraic (analytic) map. In
addition we require that $\mathcal{M}_g$ is universal in this respect. The exact definition
is a little bit technical. (If you are not interested in it just skip the next
sentences.)

For this we have to consider besides $\mathcal{M}_g$ the collection of all maps $\psi_{B,X}$ for all base varieties B and all possible families X over each B. If we fix a base variety B we get a map which assigns to every isomorphy class of families over B an algebraic map from B to $\mathcal{M}_g$. We call this assignment ψ_B. If we vary B we get a map ψ assigning to every B the ψ_B. If $\pi : B' \to B$ is an algebraic map and X is a family over B, then the *pullback* π^* is a family over B'. (Roughly speaking, for defining the pullback we take the fibre $X_{\pi(b)}$ over the basepoint b to get a family over B'.)

Now we have the following situation:

$$\begin{array}{ccc} B' & \xrightarrow{\psi_{B',\pi^*X}} & \mathcal{M}_g \\ \downarrow{\scriptstyle \pi} & & \| \\ B & \xrightarrow{\psi_{B,X}} & \mathcal{M}_g \end{array} \qquad (7.1)$$

and we see that the diagram commutes:

$$\psi_{B',\pi^*X} = \psi_{B,X} \circ \pi.$$

Now let N be another variety and ϕ a rule of assignment similar to ψ. ϕ defines for every variety B a map ϕ_B, which assigns to every isomorphy class of families over B an algebraic map $\phi_{B,X}$ from B to N. We require in addition for this assignment that the corresponding diagram (7.1) commutes, i.e.

$$\phi_{B',\pi^*X} = \phi_{B,X} \circ \pi.$$

If we take for B just a point $\{*\}$ then the possible families are exactly the isomorphy classes of curves. Hence every isomorphy class defines a point in N and we get a map of sets

$$\phi_{\{*\}} : \mathcal{M}_g \to N.$$

($\mathcal{M}_g$ is by definition the set of isomorphy classes of curves.) But $\mathcal{M}_g$ and N are algebraic varieties. For $\mathcal{M}_g$ to be universal we require that this is an algebraic map for all possible pairs (N, ϕ).

$\mathcal{M}_g$ together with its structure as a variety is called a *coarse moduli space*. There is an even better concept. It would be highly favourable if there exists a family $\mathcal{U}$ of curves over $\mathcal{M}_g$

$$\pi : \quad \mathcal{U} \quad \to \quad \mathcal{M}_g$$

such that the fibre over a point $p \in \mathcal{M}_g$ is a representative for the class p and every family of curves X over B is induced by pulling back $\mathcal{U}$ via the map $\psi_{B,X}$. The situation is exhibited by the diagram

$$\begin{array}{ccccc} X \cong \psi_{B,X}^*(\mathcal{U}) & \longrightarrow & & & \mathcal{U} \\ \downarrow & & \downarrow & & \downarrow{\scriptstyle \pi} \\ B = & & B & \xrightarrow{\psi_{B,X}} & \mathcal{M}_g \end{array}$$

In this case we would call $\mathcal{U}$ a *universal family* and $\mathcal{M}_g$ a *fine moduli space*. Unfortunately $\mathcal{M}_g$ is never a fine moduli space. But

Theorem 7.2. *(a) $\mathcal{M}_g$ exists as a coarse moduli space and is a quasi-projective variety.*

(b) For genus $g > 2$ we can find an open dense subset of $\mathcal{M}_g$ where there exists a universal family.

The obstruction to the existence of a universal family is the existence of curves which have nontrivial automorphisms. Now a generic curve of genus $g > 2$ has no automorphism besides the identity.[1] The notation "generic" is shorthand for the fact that the isomorphism classes of curves which have nontrivial automorphisms are contained in a closed set of codimension 1 in $\mathcal{M}_g$.

7.2 Methods of Construction

$\mathcal{M}_0$ is just one point. There are different ways to see this. For example let X be a Riemann surface of genus 0 and $\mathbb{P}^1$ be the standard projective line. We choose a point $p \in X$. Because

$$\deg[p] = 1 \geq 2g - 1$$

we calculate by Riemann–Roch

$$l([p]) = \deg[p] + 1 - g = 2.$$

This says there exists a function f having exactly a pole of multiplicity 1 at the point p and which is holomorphic elsewhere. In other words the map $X \to \mathbb{P}^1$ defined by f is 1:1 and surjective, hence an analytic isomorphism.

$\mathcal{M}_1$. Let us first consider the analytic way. Let T' be an arbitrary torus. In Sect. 3.3 we saw that we can always find a torus T in the same isomorphy class described by

$$T = \frac{\mathbb{C}}{L}, \qquad L = \mathbb{Z} \oplus \tau\mathbb{Z}, \quad \tau \in \mathcal{H}.$$

Up to a global coordinate change in $\mathbb{C}$ the lattice fixes the isomorphy class. Of course we can still change the basis in our lattice. Such a coordinate change is necessarily given by a matrix

$$M = \begin{pmatrix} a & b \\ c & d \end{pmatrix} \in \mathrm{Sl}\,(2, \mathbb{Z}).$$

After normalizing again we see that the lattice

$$L' = \mathbb{Z} \oplus \tau'\mathbb{Z}$$

with

---

[1] See Griffiths/Harris, [GH], Principles of Algebraic Geometry, p. 276.

$$\tau' = M \cdot \tau = \frac{a\tau + b}{c\tau + d} \qquad (7.2)$$

yields exactly the same isomorphy class of tori as T. The $\tau \in \mathcal{H}$ which are inequivalent under the action (7.2) of $\mathrm{Sl}(2, \mathbb{Z})$ parameterize the set of our one-dimensional tori. If we examine the generators of the group of operations we can fix a fundamental region F for the group action by

$$F := \left\{ z \in \mathcal{H} \mid -\frac{1}{2} \le \mathrm{Re}\, z < \frac{1}{2}, \quad \begin{cases} |z| \ge 1 & \text{if } \mathrm{Re}\, z \le 0 \\ |z| > 1 & \text{if } \mathrm{Re}\, z > 0 \end{cases} \right\}.$$

Every orbit has exactly one representative in F (see Fig. 7.1).

In higher dimensions we had to introduce the concept of polarization and use the symplectic group $\mathrm{Sp}(2g, \mathbb{Z})$. This is not necessary here because as the reader can easily calculate

$$\mathrm{Sp}(2, \mathbb{Z}) = \mathrm{Sl}(2, \mathbb{Z}).$$

Hence the set of isomorphy classes of one-dimensional tori is the same as the set of isomorphy classes of principally polarized one-dimensional tori.

We can also give this quotient the structure of a noncompact Riemann surface by using the j function already defined in Sect. 3.3 by

$$j(z) := 1728 \cdot \frac{g_2{}^3(z)}{\Delta(z)}.$$

This j is called the *modular function*. It obeys

$$j(M \cdot z) = j(z), \quad M \in \mathrm{Sl}(2, \mathbb{Z}).$$

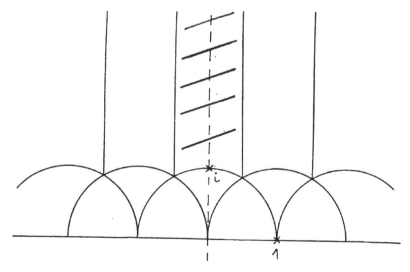

Fig. 7.1. The fundamental domain

It is holomorphic in $\mathcal{H}$ and has a so-called q-expansion for z with big imaginary part:

$$j(z) = \frac{1}{q} + \sum_{n=0}^{\infty} c_n q^n, \qquad c_n \in \mathbb{Z} \quad q = \exp(2\pi i\, z).$$

Now j takes every value of $\mathbb{C}$ and takes it just once if we restrict the argument to F. Hence we are able to define a complex structure on the quotient

$$\mathcal{H}/\mathrm{Sl}\,(2, \mathbb{Z})$$

via j and get the result

$$\mathcal{M}_1 \cong \mathbb{C}.$$

Of course it is now necessary to check whether the complex structure has really the required properties for a moduli space. We will skip this. Let us only remark that it is a coarse moduli space but not a fine one.

The fundamental idea in this analytic approach was to use the universal covering of the torus. We can also work completely in the algebraic category without using this covering. Here we use the fact that elliptic curves (the algebraic analogues of the tori) can be represented as nonsingular cubic curves in $\mathbb{P}^2$. After a coordinate change we can assume the curve $\mathcal{E}_\lambda$ to be given by

$$\mathbf{Z}\mathbf{Y}^2 = \mathbf{X}(\mathbf{X} - \mathbf{Z})(\mathbf{X} - \lambda\mathbf{Z}), \qquad \lambda \neq 0, 1. \tag{7.3}$$

There is still some choice involved in interchanging

$$(0, 1, \lambda)$$

by an element of the group $\mathcal{S}_3$ (the permutation group of three elements) and transforming the result back to the form $(0, 1, \lambda')$ by a fractional linear transformation of $\mathbb{P}^1$. This is due to the fact that we can interchange the role played by the elements $(0, 1, \lambda)$. As a result we get

$$\mathcal{E}_\lambda \cong \mathcal{E}_{\lambda'}$$

if and only if

$$\lambda' \in \left\{ \lambda, \frac{1}{\lambda}, 1 - \lambda, \frac{1}{1 - \lambda}, \frac{\lambda}{\lambda - 1}, \frac{\lambda - 1}{\lambda} \right\}.$$

Hence the moduli space $\mathcal{M}_1$ is given as a quotient

$$\mathbb{C} \setminus \{0, 1\}$$

by the above action of $\mathcal{S}_3$. Here again

$$j(\lambda) = 2^8 \cdot \frac{(\lambda^2 - \lambda + 1)^3}{\lambda^2(\lambda - 1)^2} \tag{7.4}$$

is invariant under this action and gives an isomorphism

$$\mathcal{M}_1 \to \mathbb{C}.$$

In fact both j functions are the same.

$\mathcal{M}_g$, $g \geq 2$. Different attacks are possible.

Using covering space. Let X be a Riemann surface of genus $g \geq 2$. The universal covering space of X is the upper half-plane $\mathcal{H}$. The group of analytic automorphism of $\mathcal{H}$ is the group

$$\mathrm{PSL}(2,\mathbb{R})$$

which is the group of all *fractional linear transformations* with real coefficients

$$z \mapsto M \cdot z = \frac{az+b}{cz+d}, \qquad M = \begin{pmatrix} a & b \\ c & d \end{pmatrix} \in \mathrm{Sl}(2,\mathbb{R}). \tag{7.5}$$

The group of covering transformations of $\mathcal{H}$ over X is the fundamental group $\pi_1(X)$, see Sect. 2.3. It is a discrete subgroup of $\mathrm{PSL}(2,\mathbb{R})$ and consequently X is given as the quotient

$$X = \mathcal{H}/\pi_1(X).$$

Essentially the different isomorphy classes of X can be described by the different classes of subgroups. This is the same situation as in the case of the tori above. There $\pi_1(T)$ was the lattice and the covering space was $\mathbb{C}$. The moduli of the tori was transferred to the moduli of lattices.

Using Teichmüller space. We fix a Riemann surface X_0. A *Teichmüller surface* is an arbitrary Riemann surface X together with a topological isomorphism $f : X \to X_0$. We define an equivalence relation

$$(X,f) \sim (X',f')$$

if there exists a topological isomorphism $\psi : X \to X'$ such that

$$g := f' \circ \psi \circ f^{-1} : \ X_0 \to X_0$$

$$\begin{array}{ccc} X & \xrightarrow{f} & X_0 \\ \downarrow{\psi} & & \downarrow{g} \\ X' & \xrightarrow{f'} & X_0 \end{array}$$

is the identity up to isotopy. An *isotopy* is a homotopy where all intermediate maps are also topological isomorphisms.

Teichmüller space $\mathcal{T}_g$ is the set of equivalence classes of Teichmüller surfaces. It can be given in a natural way a geometric structure as an open contractible subset of $\mathbb{C}^{3g-3}$. Now

$$\mathcal{M}_g = \mathcal{T}_g/\Gamma_g$$

where Γ_g is the so-called *mapping class group*

$$\Gamma_g = \frac{\mathrm{Diff}^+(X_0)}{\mathrm{Diff}_0(X_0)}.$$

$\mathrm{Diff}^+(X_0)$ is the group of all diffeomorphisms compatible with the orientation and $\mathrm{Diff}_0(X_0)$ is the subgroup of diffeomorphisms which are isotopic to the identity.

Using Torelli's theorem. Via the Torelli correspondence we can shift the moduli problem of curves of genus g to the moduli problem of principally polarized abelian varieties of dimension g. In Chap. 5 we saw that the set of isomorphy classes of the latter is given by the quotient of Siegel's upper half-space by the symplectic group $\mathrm{Sp}\,(2g, \mathbb{Z})$. Let us recall this action. On

$$\mathcal{H}_g = \{\ \Pi \in \mathrm{Mat}\,(g, \mathbb{C}) \mid \quad \Pi = {}^t\Pi,\ \mathrm{Im}\,\Pi > 0\ \}$$

the matrix

$$M = \begin{pmatrix} A\ B \\ C\ D \end{pmatrix} \in \mathrm{Sp}\,(2g, \mathbb{Z})$$

operates by

$$\Pi' = M \cdot \Pi = (A \cdot \Pi + B)(C \cdot \Pi + D)^{-1}.$$

We denote the quotient space by $\mathcal{A}_g$. Now Torelli embeds $\mathcal{M}_g$ into $\mathcal{A}_g$ by assigning to every curve its Jacobian $\mathrm{Jac}\,(X)$. In fact $\mathcal{A}_g$ is a quasi-projective variety and $\mathcal{M}_g$ is a subvariety. The problem is how to describe the principally polarized abelian varieties arising as Jacobians explicitly. This problem is famous under the name Schottky problem. There are some kinds of solutions, none of them completely satisfying.

There is another way to construct the moduli space by **_purely algebraic techniques_**. I do not want to go into details here. The main idea is to embed the curve via m-canonical forms (also called m-differentials) into a fixed projective space and factor out some suitable group action.

7.3 The Geometry of the Moduli Space and Its Compactification

Now the whole algebraic geometric technique is available to study these moduli spaces. We only want to quote some results.

Proposition 7.3.
(a)

$$\dim \mathcal{M}_g = \begin{cases} 0, & g = 0 \\ 1, & g = 1 \\ 3g - 3, & g \geq 2. \end{cases}$$

(b) $\mathcal{M}_g$ *is irreducible.*

The assertion of the dimension comes from the following argument. Let $p \in \mathcal{M}_g$ be a nonsingular point, C the corresponding curve (or Riemann surface), T_C be the holomorphic tangent line bundle and ω_C be its dual the holomorphic cotangent bundle. Then the tangent space of $\mathcal{M}_g$ at the point p is given by [2]

$$H^1(C, T_C).$$

By Serre duality (see (8.11))

$$H^1(C, T_C) \cong H^0(C, \omega_C \otimes T_C^*)^* = H^0(C, \omega_C^2)^*. \qquad (7.6)$$

The bundle ω_C corresponds to the canonical divisor class K and ω_C^2 corresponds to $2K$ (in Chap. 9 we will see this), hence

$$H^0(C, \omega_C^2) \cong L(2K).$$

The sections of the line bundle ω_C are the holomorphic differentials, and the sections of the bundle ω_C^2 are the *quadratic differentials*. Locally a basis of them is given by

$$dz \otimes dz.$$

For $\mathbf{g} = \mathbf{0}$ we obtain $\deg(2K) = -4$ and hence $L(2K) = \{0\}$.

For $\mathbf{g} = \mathbf{1}$ we have $\omega_C = dz$. In particular, the canonical class and all its powers are trivial. This implies $\dim L(2K) = 1$.

For $\mathbf{g} = \mathbf{2}$ we have

$$\deg(2K) = 4g - 4 \geq 2g - 1$$

and hence by Riemann–Roch

$$\dim L(2K) = 4g - 4 - g + 1 = 3g - 3.$$

Hence the assertion about the dimension.

Everything we said above at the point p extends to the local situation in the moduli space. This means we can say that the tangent space of $\mathcal{M}_g$ at the point p is given in a natural way by the dual space of the globally defined quadratic differentials on the curve C represented by p.

The quasi-projective variety $\mathcal{M}_g$ has singularities. The reason for the singularities is again the existence of nontrivial automorphisms. For $p \in \mathcal{M}_g$ to be a singular point it is necessary for the curve C represented by p to have a nontrivial automorphism. For genus $g \geq 4$ this is also sufficient.

To apply fully the algebraic geometry we should look for a projective variety $\overline{\mathcal{M}_g}$ for which our $\mathcal{M}_g$ is an open subvariety. Such a variety is called a *compactification*. One first method is the *Satake compactification*. There is a well-developed technique to compactify $\mathcal{A}_g$. If we take the closure of $\mathcal{M}_g$ in $\overline{\mathcal{A}_g}$ we get a compactification $\overline{\mathcal{M}_g}$. Now the *boundary component*

$$\overline{\mathcal{M}_g} \setminus \mathcal{M}_g$$

---

[2] In the analytic context we are able to identify the deformation of the complex structure with Čech 1-cocycles with values in the tangent sheaf (see Chap. 8 for the definition of Čech cocycles).

has codimension 2 in $\overline{\mathcal{M}_g}$ if $g \geq 3$. In general the boundary will have singularities. It is possible to show that they are not too bad. In technical terms the compactified moduli space is still a *normal* variety. We do not want to define "normality" here. The only important fact for us is that the zeroes and poles of a function are given by 1-codimensional subvarieties. This has an interesting consequence

Proposition 7.4. *On $\mathcal{M}_g$ for $g \geq 3$ there are no nonconstant holomorphic functions.*

Proof. Assume there is such a function f. Because the boundary has codimension 2 the function f can be extended to $\overline{\mathcal{M}_g}$. But this variety is projective and hence f has to have some poles. Because the set of poles has codimension 1 it has to meet $\mathcal{M}_g$ which contradicts our starting point.   □

This compactification is unsatisfactory because one should look for a moduli problem which yields in a natural way a moduli space which is already projective. One such way is also to allow families with singular curves. The problem now is that for this moduli problem there is no "nice" moduli space. Hard algebraic geometry (the so-called *geometric invariant theory*) says which classes of curves one has to allow to get a natural projective $\overline{\mathcal{M}_g}$. These are the so-called *stable curves*. A stable curve is a connected curve which has only nodes as singularities. The nodes are sometimes called double points. Nodes look locally like the intersection of two lines. It is allowed for a stable curve to have different components, but there is a technical restriction on the intersection of the components.[3] There is also a way to define a genus for such stable curves. Of course we restrict our families to stable curves of a fixed genus.

   This $\overline{\mathcal{M}_g}$ is different from the one above. It is called *Mumford–Deligne–Knudsen* compactification. The boundary

$$\Delta = \overline{\mathcal{M}_g} \setminus \mathcal{M}_g$$

has codimension 1 and decomposes into the irreducible components (each of codimension 1)

$$\Delta = \bigcup_i \Delta_i, \qquad i = 0, 1, \ldots, \left[\frac{g}{2}\right].$$

Here Δ_0 is the closure of the set of irreducible singular curves with exactly one node as singularity (see Fig. 7.2).

   Δ_i is the closure of the set of stable curves which are the union of a nonsingular curve of genus i and one of genus $g - i$ connected at one common point (see Fig. 7.3).

   Of course there are stable curves with more nodes, but they lie in codimension 2 and higher. In Figs. 7.2 and 7.3 the picture on the right shows the

---

[3] If a nonsingular component D of C is isomorphic to $\mathbb{P}^1$ then this component has to meet the rest of the curve in at least three different points.

Fig. 7.2. Δ_0

topological situation, the picture on the left indicates the typical situation if we take a "suitable" real part of the complex curve.

In the case of genus 1 one expects the compactification of $\mathcal{M}_1 \cong \mathbb{C}^1$ to be $\mathbb{P}^1$. In fact if we set $\lambda = 0$ in the defining polynomial (7.3) of the elliptic curve and replace $\mathbf{X}$ by $\mathbf{X} + \mathbf{Z}$ (which is a projective transformation) we get the *nodal cubic* $\mathcal{N}$

$$\mathbf{ZY}^2 = \mathbf{X}^2(\mathbf{X} + \mathbf{Z}).$$

It has a singular point at $(0, 0, 1)$ which is a node. $\mathcal{N}$ is hence a stable curve. In Fig. 7.4 the real part of the complex curve obtained by allowing only real numbers as coordinates is given.

The j-function (7.4) again gives a parameterization because $\lambda = 0$ is a pole of j. The other pole $\lambda = 1$ yields another singular curve which is equivalent to the above by projective linear transformation. $\mathcal{N}$ is the only stable genus 1 curve besides the nonsingular ones.

There is another irreducible and singular cubic. It is given by

$$\mathbf{ZY}^2 = \mathbf{X}^3.$$

Figure. 7.5 shows its real part.

The singular point is again $(0, 0, 1)$. But now it is the so-called *cusp*, hence the curve is not stable . The curve is called a *cuspidal cubic*.

The fact that this curve is not included in the compactified moduli space does not mean that it cannot occur together with the elliptic curves in a family. For example $(a, b \in \mathbb{C})$

$$
\begin{aligned}
\mathcal{E}_1 \quad &: \quad \mathbf{ZY}^2 = \mathbf{X}^3 + a\mathbf{XZ}^2 + b\mathbf{Z}^3 \quad \text{and} \\
\mathcal{E}_\mu \quad &: \quad \mathbf{ZY}^2 = \mathbf{X}^3 + a\mu^2\mathbf{XZ}^2 + b\mu^3\mathbf{Z}^3
\end{aligned}
$$

with $\mu \in \mathbb{C}, \mu \neq 0$ have the same value of the j-function and hence are isomorphic. Now by allowing μ to vary over $\mathbb{C}$ we get a family of cubic curves $\mathcal{E}_\mu$ where

Fig. 7.3. Δ_i

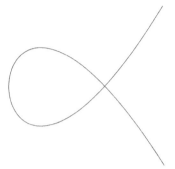

Fig. 7.4. Nodal Cubic

$$\mathcal{E}_\mu \cong \mathcal{E}_\nu \quad \text{if} \quad \mu, \nu \neq 0$$

and $\mathcal{E}_0$ is the cuspidal cubic. This also shows that by allowing this curve to be a member of the families we would get rather wild "moduli spaces".

Let us return to the general situation. What are the linear equivalence classes of the divisors in $\mathcal{M}_g$ or $\overline{\mathcal{M}_g}$, in other words what is $\operatorname{Pic} \mathcal{M}_g$ and $\operatorname{Pic} \overline{\mathcal{M}_g}$. Remember we defined the group Pic as the group of divisors modulo divisors coming from functions. The knowledge of Pic gives us information on the existence of certain objects (line bundles, sections of line bundles and so on) with prescribed properties at certain subsets of points in the moduli space. In string theory one is looking for exactly such objects. One of the most important divisor classes is the *Hodge class* λ. It is defined in the following way (see also Chaps. 9 and 10): Let p be a point in the moduli space, C the corresponding curve. At the point p we have the vector space of global holomorphic differentials on the curve C. These vector spaces can be glued together to a rank g vector bundle. Taking the highest exterior power (the determinant) we get a line bundle on $\mathcal{M}_g$. This bundle is called the *Hodge*

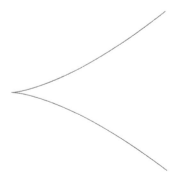

Fig. 7.5. Cuspidal Cubic

bundle. It can be shown that there is a well-defined correspondence assigning to every line bundle a divisor class.[4] We denote this divisor class also by λ.

Proposition 7.5. *Let* $g \geq 3$ *then*

$$\operatorname{Pic} \mathcal{M}_g \cong \mathbb{Z} \cdot \lambda.$$

In other words, all nontrivial divisor classes are multiples of the Hodge class.

The situation gets more complicated if we consider $\overline{\mathcal{M}}_g$. On $\overline{\mathcal{M}}_g$ we have the components of the boundary

$$\Delta_i, \quad i = 0, 1, \ldots, \left[\frac{g}{2}\right].$$

These are divisors. We want to define the Hodge class on the whole $\overline{\mathcal{M}}_g$. Unfortunately not λ itself but only 2λ admits a natural extension to $\overline{\mathcal{M}}_g$. We set δ_i for the linear equivalence class of Δ_i for $i = 0, 2, 3, \ldots, [g/2]$. We set (in a formal sense) δ_1 equal to $1/2$ of the linear equivalence class of Δ_1, and (again in a formal sense)

$$\hat{\lambda} = \frac{1}{2} \cdot (2\lambda).$$

Because $\hat{\lambda} = \lambda$ on $\mathcal{M}_g$ we drop the hat again. We consider the $\mathbb{Q}$-vector space W which is generated freely by the above elements:

$$W = < \ \delta_0, \delta_1, \delta_2, \ldots, \delta_{[\frac{g}{2}]}, \lambda > .$$

Proposition 7.6. [5] *Let* $g \geq 3$ *then* $\operatorname{Pic} \overline{\mathcal{M}}_g$ *is a lattice in the* $\mathbb{Q}$*-vector space* W *of full rank. More precisely it is freely generated over* $\mathbb{Z}$ *by the divisor classes*

$$2\lambda, \lambda + \delta_1, \delta_0, \delta_2, \ldots, \delta_{[\frac{g}{2}]}.$$

Let me make some remarks:

(1) The definitions of δ_1 and λ for constructing the vector space W may look very ad hoc. But there is a deeper reason behind this. One is able to define the *Picard group* of the "*moduli functor*". This group is a free abelian group generated by the above elements. Now $\operatorname{Pic} \overline{\mathcal{M}}_g$ is the above subgroup of index 2 in it.

(2) One often defines the Picard group as the group of isomorphy classes of line bundles (see Chap. 9). In the case of nonsingular projective varieties both definitions coincide. Unfortunately $\mathcal{M}_g$ and $\overline{\mathcal{M}}_g$ have singular points. Hence the quoted result on $\operatorname{Pic} \overline{\mathcal{M}}_g$ is not valid with respect to the other definition. The only thing we know here is that this group is again a subgroup of finite index of the Picard group of the moduli functor.

---

[4] Essentially take the divisor of a meromorphic section of the bundle, see Chap. 9
[5] Arbarello,Cornalba,[AC], Prop.2

(3) You may ask why this speciality of δ_1. As in the existence problem of the universal family, the important fact here is again that even a generic curve corresponding to a point of Δ_1 has a nontrivial automorphism. Δ_1 consists of the *curves with elliptic tails*. Even if we fix one point p (the connection point to the rest of the curve) then there is always the automorphism $x \mapsto -x$ on the elliptic tail. Here the $-$ is with respect to the group structure on the elliptic curve with p as zero element. This extends to a nontrivial automorphism of the whole curve, by trivial operation on the rest.

In the description of the moduli space via principally polarized abelian varieties we have the technique of modular forms available. Let us assume for the following $g \geq 2$ which is anyway the only interesting case. We call f a *modular form* of weight m if

(1) f is holomorphic on $\mathcal{H}_g$ and
(2)

$$
f(M \cdot \Pi) = f\left(\frac{A \cdot \Pi + B}{C \cdot \Pi + D}\right) = (\det(C \cdot \Pi + D))^m \cdot f(\Pi),
$$

$$
M = \begin{pmatrix} A & B \\ C & D \end{pmatrix} \in \mathrm{Sp}\,(2g, \mathbb{Z}).
$$

(7.7)

Of course f does not define a function, neither on the quotient space $\mathcal{A}_g$ nor on the subvariety of the Jacobians $\mathcal{J}_g$. In fact it defines a section of the mth power of the Hodge line bundle.

Modular forms of sufficient high weights can be used to prove the quasi-projectivity of $\mathcal{A}_g$ by embedding it into some $\mathbb{P}^n$ just as we embedded the principally polarized abelian varieties by theta functions. Even more astonishing, theta functions will do the job here also. As remarked in Sect. 6.3, $\vartheta(z, \Pi)$ is a holomorphic function on $\mathbb{C}^g \times \mathcal{H}_g$. By restricting ourselves to $z = 0$ (*theta null values*) and a certain subgroup Γ of $\mathrm{Sp}\,(2g, \mathbb{Z})$ of finite index we get

$$
\vartheta(0, M \cdot \Pi) = \chi_M \cdot (\det(C\Pi + D))^{1/2} \cdot \vartheta(0, \Pi)
$$

with

$$
M \in \Gamma, \qquad \chi_M \in \mathbb{C}, \qquad \chi_M{}^8 = 1.
$$

So roughly speaking, theta null values are modular forms of weight $1/2$. To get honest modular forms under the whole of $\mathrm{Sp}\,(2g, \mathbb{Z})$ we have to take suitable polynomials of the theta functions with characteristics.

In view of some recent speculations on p-adic or even adelic string theory let me make a final remark. Up to now we have just used algebraic geometric language to describe more or less the same objects as can be given analytically. But the validity of algebraic geometry is much broader than this. We can work over arbitrary fields whether they are algebraically closed (like $\mathbb{C}$) or not (like $\mathbb{Q}$) and even over rather general commutative rings. Hence we are also able to include "arithmetic features". For example the moduli space of curves comes with such an arithmetic structure. We can look for curves which after a suitable coordinate change are given by a polynomial with rational (or equivalently integer) coefficients. These polynomials have arithmetic properties. For example we can ask whether they are still nonsingular if we reduce them modulo a prime number. Or we can ask whether there are integer solutions to the polynomials (the so-called *rational* points on the curve). There is a well-developed theory of finite field extensions of $\mathbb{Q}$. These fields are called number fields. They have arithmetic properties similar to those of $\mathbb{Q}$. For example they also contain the so-called ring of integers.

Now the isomorphy classes of curves which can be defined as polynomials with coefficients from a number field (depending on the curve) form a dense subset of the moduli space $\mathcal{M}_g$. In this sense $\mathcal{M}_g$ carries an arithmetic structure.

Hints for Further Reading

There is a nice survey article

[H]   Harris, J., An Introduction to the Moduli Space of Curves (in) *Mathematical Aspects of String Theory, Proceed. 1986*, ed. Yau, pp. 285–312, World Scientific, Singapore, 1987.

This is a quite readable collection of results on the moduli space of curves.

For the moduli of elliptic curves see also

[MS]   Mumford, D., Suominen, K., Introduction to the Theory of Moduli (in) *Algebraic Geometry Oslo 1970*, Wolters-Nordhoff, Groningen, 1971.

Unfortunately, more or less all texts on the subject which give not only just an overview but also proofs are very difficult to read. All of them use rather advanced algebraic geometry. Here I list only two of them:

[GIT] Mumford, D., *Geometric Invariant Theory,* 2nd edition, Springer, 1982

is the fundamental text. In the appendices you find most of the important results.

[M]   Mumford, D., Stability of Projective Varieties, *L'Enseign. Math.* **23** (1977) 39–110.

Here you find the analysis of which singular curves one should admit in the compactification of the moduli space.

[GH]   Griffiths, Ph., Harris, J., Principles of Algebraic Geometry, Wiley, 1978 can be used for general background information.

If you want to learn more on divisors on the moduli space, especially which divisor classes contain positive divisors, you should consult as an introduction Harris article [H] mentioned above. Unfortunately he covers the subtlety of the Picard group of $\overline{\mathcal{M}_g}$ in a slightly oversimplifying way. You can find precise results in

[AC]   Arbarello, E., Cornalba, M., The Picard Groups of the Moduli Space of Curves, *Topology* **26** (1987) 153–171.

To study this paper you will also need the paper of Mumford on stability [M] and

[HM] Harris, J., Mumford, D., On the Kodaira Dimension of the Moduli Space of Curves, *Invent. math.* **67** (1982) 23–86.

The singularities of the moduli space are treated in

[Po]   Popp, H., The Singularities of the Moduli Schemes of Curves, *J. Number Theory* **1** (1969) 90–107,

in algebraic geometric language. In analytic language you find results in

[R]   Rauch, H.E., Singularities of Modulus Space, *Bull. Am. Math. Soc.* **68** (1962) 390–394.

For an approach via Teichmüller theory you may check the extensive list of literature in

[Be]   Bers, L., Finite Dimensional Teichmüller Spaces and Generalizations, *Bull. Am. Math. Soc.* **5** (1981) 131–172.

It also serves as a good introduction to the subject.
There is another recent book

[L]   Lehto, O., *Univalent Functions and Teichmüller Spaces*, Springer, 1986,

which contains a lot of interesting material.

8

Vector Bundles, Sheaves and Cohomology

8.1 Vector Bundles

For the beginning let M be a differentiable manifold. Before giving the exact definition of a vector bundle let me start with an example. We introduced in Chap. 4 the tangent space at a point $x \in M$. If we vary the point x on the manifold we will get a "family of vector spaces" by assigning to every x its tangent space. The dimension of each vector space will be the same and the tangent space at nearby points "change continuously" from one to the other. Intuitively this is exactly what we understand by a vector bundle. Hence we call this family of all tangent spaces the *tangent bundle* of the manifold M.

Now we come to the exact definition. Let F and M be differentiable manifolds,

$$\pi : F \to M$$

a differentiable map, then F together with the map π is called a *family of $\mathbb{R}$-vector spaces* if every fibre

$$\pi^{-1}(x), \quad x \in M$$

of the map π is a real vector space. If we have two families

$$\pi : F \to M, \qquad \rho : F' \to M$$

over the same manifold M, then a *homomorphism of families ψ* from F to F' is a differentiable map $\psi : F \to F'$ which respects the fibres, i.e.

$$\rho(\psi(z)) = \pi(z), \quad z \in F$$

and which is a linear map when restricted to each single fibre. There exist for every M the trivial family

$$M \times \mathbb{R}^k \to M, \quad k = 0, 1, 2, \dots .$$

M. Schlichenmaier, Vector Bundles, Sheaves and Cohomology. In: M. Schlichenmaier, An Introduction to Riemann Surfaces, Algebraic Curves and Moduli Spaces, Theoretical and Mathematical Physics, 87–101 (2007)
DOI 10.1007/978-3-540-71175-9_9

If $\pi : F \to M$ is a family over M, then for every open set U in M the restriction

$$\pi_| : \pi^{-1}(U) \to U$$

is a family over U. Instead of this the notation $F_{|U}$ is commonly used. In most cases we will also drop explicit mention of the map π.

Definition 8.1. *A vector bundle of rank k is a family of k-dimensional vector spaces which is locally trivial, which means: for all x in M we have an open set U in M containing x such that*

$$E_{|U} \cong U \times \mathbb{R}^k$$

in the sense of homomorphism of families of vector spaces.
A vector bundle of rank 1 is called a line bundle.

Let us fix a vector bundle F of rank k. It is clear from the definition that we can always cover M by open sets U_i such that $F_{|U_i}$ is trivial. Such a collection

$$\mathcal{U} = (U_i)_i \in I$$

is called a *trivializing covering*. In the case that M is compact we can achieve this with finitely many U_i.
Now let U_i and U_j be members of such a covering with

$$U_{ji} = U_i \cap U_j \neq \emptyset.$$

Starting from the trivializing maps

$$\phi_i : \pi^{-1}(U_i) \to U_i \times \mathbb{R}^k, \qquad \phi_j : \pi^{-1}(U_j) \to U_j \times \mathbb{R}^k$$

we get the two maps

$$\phi_j, \phi_i : \pi^{-1}(U_{ji}) \to U_{ji} \times \mathbb{R}^k.$$

With these we can form the composite map

$$\phi_j \circ \phi_i^{-1}.$$

For a fixed $x \in U_{ji}$ this is a linear isomorphism from $\mathbb{R}^k$ to $\mathbb{R}^k$, hence an element of $\mathrm{Gl}(k, \mathbb{R})$. It depends in a differentiable manner on the $x \in U_i \cap U_j$ and hence it can be given by a $k \times k$ matrix C_{ji} with entries coming from the differentiable functions on U_{ji} (denoted by $\mathcal{E}(U_{ji})$). In this sense

$$C_{ji} \in \mathrm{Gl}(k, \mathcal{E}(U_{ji})).$$

In this way such a matrix C_{ji} is assigned to every ordered pair (j, i). It is easily verified that these matrices fulfil the so-called *cocycle conditions*

$$C_{ii} = 1, \quad C_{ij} = (C_{ji})^{-1},$$
$$C_{ki} = C_{kj} \cdot C_{ji} \quad \text{on} \ \ U_i \cap U_j \cap U_k.$$

This set of matrices is called a *defining cocycle* for the bundle F. The assignment of C_{ji} to the bundle F depends on the covering $\mathcal{U}$ and on the trivialization maps chosen. If we had another trivialization map ψ_i for the same U_i, then

$$U_i \times \mathbb{R}^k \xrightarrow{\phi_i^{-1}} F|_{U_i} \xrightarrow{\psi_i} U_i \times \mathbb{R}^k.$$

The composition $\psi_i \circ \phi_i^{-1}$ defines an element

$$B_i \in \mathrm{Gl}(k, \mathcal{E}(U_i)).$$

We can do this for every U_i. For the new cocycle C'_{ji} with respect to the new maps ψ_i we get

$$C'_{ji} = B_j \cdot C_{ji} \cdot B_i^{-1}.$$

If there is such a relation between two cocycle sets of matrices C'_{ji} and C_{ji} we call them *cohomologous*. It is easy to verify that being cohomologous is an equivalence relation.

 If we start with a given set C_{ji} of matrices which obey the cocycle conditions, we construct a vector bundle which has exactly these matrices as defining cocycle. This is achieved by glueing the manifolds $U_i \times \mathbb{R}^k$ along the common intersection $(U_i \cap U_j) \times \mathbb{R}^k$ using the matrices. In fact there is the following:

Proposition 8.2. *The isomorphy classes of rank k bundles which admit a fixed trivializing covering $\mathcal{U} = (U_i)$ is given by the cohomology classes of cocycles*

$$C_{ji} \in \mathrm{Gl}(k, \mathcal{E}(U_i \cap U_j)).$$

 If $\pi : F \to M$ is a vector bundle and U an open set, then we denote by $F(U)$ the set of sections over U. Here we call

$$s : U \to F$$

a *section* over U if s is a differentiable map and if it maps $x \in U$ to an element in the vector space over x (i.e. $\pi(s(x)) = x$). It is clear that we can add sections

$$(s + t)(x) := s(x) + t(x)$$

and multiply them by differentiable functions

$$(f \cdot s)(x) := f(x)s(x).$$

Hence they form a module over the ring $\mathcal{E}(U)$ (a *module* is simply a "vector space" over a ring). In particular, if $F = M \times \mathbb{R}$ is a trivial bundle, then we get $F(U) = \mathcal{E}(U)$.

 In the case that U is a trivializing open set, we can represent a section s

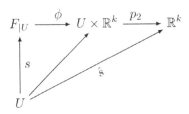

by a map

$$\hat{s} : U \to \mathbb{R}^k, \quad x \mapsto p_2(\phi(s(x))).$$

(p_2 is the projection onto $\mathbb{R}^k$.) $\hat{s}$ is a k-tuple of differentiable functions. It is defined only locally and depends on the trivialization. If a section $s \in F(U)$ is represented over $U_i \cap U_j$ by $\hat{s}_i$ and $\hat{s}_j$ then they are related by multiplication with the matrix C_{ji}. We will see later that the set of all local sections of a bundle determines the bundle.

As remarked already in Chap. 4 the sections of the tangent bundle are the vector fields and the sections of the cotangent bundle are the differentials. As you might remember, there we were able to describe them locally by functions which transform under coordinate change. But a coordinate change is just a change of the trivialization.

Starting from two vector bundles V and W we are able to define

$$V \oplus W, \quad V \otimes W, \quad V^*, \quad \wedge^l V.$$

$V \oplus W$, for example, is the fibrewise sum. If V is represented by the cocycle C_{ji}, W by the cocycle D_{ji}, then it is represented by

$$\begin{pmatrix} C_{ji} & 0 \\ 0 & D_{ji} \end{pmatrix} \in \mathrm{Gl}\,(\mathrm{rank}\,V + \mathrm{rank}\,W, \mathcal{E}(U_i \cap U_j)).$$

$V \otimes W$ is represented by the tensor product (sometimes also called Kronecker product)

$$C_{ji} \otimes D_{ji}.$$

V^* is given by

$$C_{ji}^* = ({}^t C_{ji})^{-1}.$$

The *exterior product* is given by the exterior products of the matrices C_{ji}. In the special case $\wedge^{\mathrm{rank}\,V} V$, we call this bundle $\det V$. It is a bundle of rank 1, a line bundle. Its cocycle is given by

$$\det C_{ji}.$$

The classical *tensors* are the sections of certain tensor products of the tangent and the cotangent bundle (the dual of the tangent bundle).

Let us make a final construction. For this we consider N, another differentiable manifold. We start with $f : N \to M$ a differentiable map and F a vector bundle on M. We want to define a vector bundle, the *pullback* f^*F, on

N. For every $y \in N$ we take the vector space over $f(y)$ and glue it "above" y. In terms of cocycles this means that we consider the functions in the cocycle of F via the map f as functions on the manifold N and hence get a cocycle on N which defines a bundle.

Up to now we have considered real differentiable vector bundles. It is easy to modify the above definition to define *complex, holomorphic* or *algebraic bundles*. First we consider complex vector spaces instead of real vector spaces. Now by replacing all words differentiable by complex analytic, resp. algebraic, we get holomorphic, resp. algebraic, bundles. Complex bundles make sense on real manifolds, but holomorphic bundles only on complex manifolds, resp. algebraic varieties. For example, an algebraic vector bundle of rank k is given by a covering (U_i) of Zariski open sets and cocycles C_{ji} which are invertible $k \times k$ matrices with regular functions on $U_i \cap U_j$ as entries. (By a regular function on a set U we mean rational functions without poles in this set.) If we denote the ring of regular functions on a set U by $\mathcal{O}(U)$ we see that the set of algebraic sections $F(U)$ of an algebraic bundle F is again a $\mathcal{O}(U)$ module. $\mathcal{O}(U)$ itself is the set of algebraic sections of the trivial bundle $M \times \mathbb{C}$ over U.

8.2 Sheaves

They are nothing mysterious. To start with an example, let F be a vector bundle. We assigned to every open set U an abelian group, the group of sections over U. This assignment is compatible with restricting our set U to a smaller set V. Essentially this is what makes it a sheaf. They are generalizations of vector bundles.

A sheaf $\mathcal{F}$ of abelian groups assigns to every open set U in M an abelian group $\mathcal{F}(U)$ and to every pair of open sets $V \subset U$ a homomorphism of abelian groups

$$\rho_V^U : \mathcal{F}(U) \to \mathcal{F}(V)$$

(the so-called *restriction map*) in such a way that

(1) $$\rho_U^U = id,$$

(2) $$\rho_V^U \circ \rho_U^W = \rho_V^W \quad \text{for } V \subset U \subset W .$$

Instead of $\rho_V^U(f)$ for $f \in \mathcal{F}(U)$ we use the simpler notation $f_{|V}$. In addition to these two conditions we require that for every open set U and every covering (U_i) of this open set we have
(3) if $f, g \in \mathcal{F}(U)$ with

$$f_{|U_i} = g_{|U_i}$$

for all U_i then $f = g$
(4) if a family of $f_i \in \mathcal{F}(U_i)$ is given with

$$f_{i,|U_i \cap U_j} = f_{j,|U_i \cap U_j}$$

then there also exists a $f \in \mathcal{F}(U)$ with

$$f_{|U_i} = f_i.$$

It is clear that the above-mentioned assignment of the sections of a vector bundle is a sheaf by taking the natural restriction as the map ρ_V^U . We call it the *sheaf of sections* of F.

Another important example is the *sheaf of locally constant functions with values in an abelian group* G (take $\mathbb{Z}$ as an example). We assign to U as $G(U)$ the set of locally constant functions from U to G. Of course this set is, by pointwise operation, again an abelian group. In the case that U is connected and nonempty we get

$$G(U) = G.$$

As restriction map we again take the natural restriction of functions. With this we have *sheafified* our group G. We use the same letter to denote the group and the associated sheaf.

In all the above constructions we can replace abelian groups by other algebraic objects, like vector spaces, rings, modules over a ring and modules over a sheaf of rings. If M is an algebraic variety then by assigning to a (Zariski-) open set U the set of regular functions $\mathcal{O}(U)$ on U and taking the natural restriction of functions as the restriction map we get a sheaf of rings $\mathcal{O}$, the *sheaf of regular functions*. In an analogous way we get for complex manifolds the *sheaf of holomorphic functions* $\mathcal{O}^{an}$ and for differentiable manifolds the *sheaf of differentiable functions* $\mathcal{E}$. These sheaves $\mathcal{O}, \mathcal{O}^{an}$ and $\mathcal{E}$ are also called *structure sheaves*.

A sheaf $\mathcal{F}$ is called a *sheaf of $\mathcal{O}$-modules* if $\mathcal{F}(U)$ is a module over $\mathcal{O}(U)$ for every open set U and the involved restriction maps are compatible with the module structure. Typical examples are the sections of algebraic vector bundles. In the same way we define sheaves of $\mathcal{O}^{an}$- and $\mathcal{E}$-modules.

By a *sheaf homomorphism*

$$\psi : \mathcal{F} \to \mathcal{G}$$

we understand an assignment of a homomorphism ψ_U (of abelian groups, rings and so on) to every open set U

$$\psi_U : \mathcal{F}(U) \to \mathcal{G}(U)$$

which is compatible with the restriction homomorphisms

$$
\begin{array}{ccc}
U & \mathcal{F}(U) \xrightarrow{\psi_U} \mathcal{G}(U) \\
\cup & \downarrow \qquad \downarrow \\
V & \mathcal{F}(V) \xrightarrow{\psi_V} \mathcal{G}(V)
\end{array}
$$

If we have a sequence of sheaf homomorphisms

$$\mathcal{F} \xrightarrow{\phi} \mathcal{G} \xrightarrow{\psi} \mathcal{H}$$

we say the sequence is *exact* at $\mathcal{G}$ if the following is valid:
(1) $\psi_U \circ \phi_U = 0$ for every open set U and
(2) if $\psi_U(s) = 0$ for $s \in \mathcal{G}(U)$, then we can find for every point x in U an open neighbourhood V and $r \in \mathcal{F}(V)$ with[1]

$$\phi_V(r) = s_{|V}.$$

If the sequence of sheaves

$$0 \to \mathcal{F} \to \mathcal{G} \to \mathcal{H} \to 0$$

is exact then we call it a *short exact sequence of sheaves*. Let us consider an example. For this example we use the complex setting. As above we denote by $\mathcal{O}^{an}$ the sheaf of holomorphic functions. Let $\mathcal{O}^{*,an}$ be the sheaf of nowhere vanishing holomorphic functions, i.e.

$$\mathcal{O}^{*,an}(U) := \{f \in \mathcal{O}(U) \mid f(x) \neq 0 \text{ for all } x \in U\}.$$

It is an abelian group by pointwise multiplication. Let $\mathbb{Z}$ be the sheaf of locally constant functions with values in $\mathbb{Z}$. If U is an open set we can define

$$0 \longrightarrow \mathbb{Z}(U) \xrightarrow{i_{|}} \mathcal{O}^{an}(U) \xrightarrow{e_{|}} \mathcal{O}^{*,an}(U) \longrightarrow 0.$$

Here i is the natural injection of the $\mathbb{Z}$-valued constant functions into all holomorphic functions and e is the exponential map

$$e(f) := \exp(2\pi i f), \quad \text{resp.} \quad e(f)(x) := \exp(2\pi i f(x)).$$

These maps are compatible with the restriction of the functions, hence we get a sequence of homomorphism of abelian sheaves:

$$0 \longrightarrow \mathbb{Z} \xrightarrow{i} \mathcal{O}^{an} \xrightarrow{e} \mathcal{O}^{*,an} \longrightarrow 0 \,. \tag{8.1}$$

We want to show that (8.1) is a short exact sequence of sheaves. Because $\exp(2\pi i n) = 1$ for every integer n the condition 1 is clear at every position. Let us start to show condition 2 at $\mathcal{O}^{*,an}$. Here we use the fact that if we have a nowhere vanishing holomorphic function g on a open set U we can locally at every point x find a simply connected neighbourhood V such that one branch of the logarithm of g is well defined. We set

$$f = \frac{1}{2\pi i} \log g \ \in \mathcal{O}(V).$$

---

[1] If one introduces stalks (see Wells, [We], p. 42), then this condition is the definition of stalk exactness.

This is a well-defined holomorphic function on V with $e_|(f) = g_{|V}$. Hence we have exactness at $\mathcal{O}^{*,an}$.

At $\mathcal{O}^{an}$ we see that a function $f \in \mathcal{O}(U)$ with

$$\exp(2\pi i f) = 1$$

on the whole of U has to be a fixed integer n on each connected component of U, hence $f \in i(\mathbb{Z}(U))$. Here again we have exactness. Exactness at $\mathbb{Z}$ is clear. Altogether we have shown: the sequence (8.1) is exact.

Let us now return to sheaves of $\mathcal{O}$-modules. Of course we can define sums, tensor products, dual sheaves and many more things with such $\mathcal{O}$-modules by defining them on every open set. We call a sheaf $\mathcal{F}$ of $\mathcal{O}$-modules *locally free of rank k* if every point has an open neighbourhood U such that

$$\mathcal{O}_{|U}^k \cong \mathcal{F}_{|U}.$$

Here we use $\mathcal{F}_{|U}$ to denote the restriction of $\mathcal{F}$ to the set U, which is now a sheaf on U. If we take as $\mathcal{F}$ the sheaf of sections of a vector bundle F we see that $\mathcal{F}$ is locally free of rank F. For the proof just take a local trivialization of F. We even have the important

Theorem 8.3. *Let M be an algebraic variety (or a complex manifold), then we have a 1:1 correspondence between the category of isomorphy classes of algebraic (resp. holomorphic) vector bundles of rank k and the category of isomorphy classes of locally free sheaves of $\mathcal{O}$ (resp. $\mathcal{O}^{an}$)-modules of rank k by assigning to every bundle its sheaf of sections.*

The word correspondence of categories means that in the above situation it respects all homomorphisms and constructions (sums, tensor products, dualizing and so on). This is the reason that in most of the literature there is no clear distinction between vector bundles and their sheaves of sections. In the following we will also freely jump between the two interpretations.

There is an important generalization of locally free sheaves of $\mathcal{O}$-modules. These are the *coherent sheaves*. We call a sheaf $\mathcal{F}$ of $\mathcal{O}$-modules coherent if there exists an exact sequence

$$\mathcal{O}^r \rightarrow \mathcal{O}^s \rightarrow \mathcal{F} \rightarrow 0.$$

This says that $\mathcal{F}$ can be generated locally by a finite number of elements as a module over the locally regular functions and that the set of relations under the generating elements is again finitely generated. Coherent sheaves are not necessarily locally free. Clearly, locally free sheaves are coherent.

Let me give an example of them. We take as variety $\mathbb{P}^2$ and fix a point x in $\mathbb{P}^2$. We define a sheaf $\mathcal{J}_x$ by

$$\mathcal{J}_x(U) := \{f \in \mathcal{O}(U) \mid f(x) = 0\}.$$

It is not locally free anymore, but it is coherent.[2] In addition it is a subsheaf of the structure sheaf $\mathcal{O}$. Essentially all subvarieties of a fixed algebraic variety correspond in such a way to coherent subsheaves of $\mathcal{O}$. For example, the positive divisors (1 codimensional subvarieties with multiplicities) correspond to locally free subsheaves (they are necessarily of rank 1).

8.3 Cohomology

Mainly we are interested in the global sections of certain sheaves. These are, for example, globally defined differentials, vector fields and so on. The problem is that even starting from a short exact sequence of sheaves like (8.1)

$$0 \longrightarrow \mathbb{Z} \xrightarrow{i} \mathcal{O}^{an} \xrightarrow{e} \mathcal{O}^{*,an} \longrightarrow 0$$

we cannot conclude that the sequence of global sections

$$0 \longrightarrow \mathbb{Z}(M) \xrightarrow{i} \mathcal{O}^{an}(M) \xrightarrow{e} \mathcal{O}^{*,an}(M) \longrightarrow 0 \qquad (8.2)$$

is exact. In fact it is not even true for general M. A measure of how far (8.2) is away from the exactness is the cohomology.

Sheaf cohomology theory is a very well developed mathematical machinery. For a fixed variety (or manifold) it is a rule to assign in a natural way to every sheaf $\mathcal{F}$ of abelian groups a sequence of abelian groups

$$\mathrm{H}^k(M, \mathcal{F}), \quad k \in \mathbb{Z}, k \geq 0.$$

The theoretical starting point is to interpret this assignment as constructing the "right derived functors of the global section functor". To develop this here would be too space consuming.[3] Instead let me take another approach using Čech cocycles.

In Chap. 6 we defined what projective varieties are. For the Čech cohomology we also have to use affine varieties. They are defined in a completely analogous manner. We start with $\mathbb{C}^n$ and define an *affine variety* as a common zero set of a finite number of polynomials in n variables. All algebraic (projective, quasi-projective) varieties can be covered by open sets which are algebraically isomorphic to affine varieties. For $\mathbb{P}^n$ we had the standard covering (see Sect. 1.2)

$$U_i := \{(z_0 : z_1 : \cdots : z_n) \mid z_i \neq 0\}, \quad i = 0, 1, \ldots, n$$

and maps

$$\varphi_i : U_i \to \mathbb{C}^n, \quad (z_0 : z_1 : \cdots : z_n) \mapsto \left(\frac{z_0}{z_i}, \frac{z_1}{z_i}, \ldots, \frac{z_{i-1}}{z_i}, \frac{z_{i+1}}{z_i}, \ldots, \frac{z_n}{z_i}\right).$$

---

[2] For a similar example in the analytic context see Wells, [We], p. 41.
[3] See Wells, [We], p. 51.

As you see these are algebraic isomorphisms, because they are given by the quotient of polynomials with nonvanishing denominator.

For the sake of simplicity we assume the existence of a finite affine covering (a countable one would also do) in the following. In particular, this is valid for all projective varieties. Let $\mathcal{U} = (U_i)_{i=1,\dots,n}$ be an affine covering of M. If

$$\sigma = (i_0, i_1, i_2, \dots, i_p)$$

is a multi-index then we set

$$U_\sigma = U_{i_0} \cap U_{i_1} \cap U_{i_2} \cdots \cap U_{i_p}$$

for the corresponding intersection of the covering sets. We define the p-cochain group by

$$C^p(\mathcal{U}, \mathcal{F}) := \prod_\sigma \mathcal{F}(U_\sigma)$$

where the product runs over all multi-indices of length $p+1$ which obey the condition of increasing index numbers

$$\sigma = (i_0, i_1, i_2, \dots, i_p) \quad \text{with } i_0 < i_1 < i_2 < \dots < i_p.$$

For example

$$C^0(\mathcal{U}, \mathcal{F}) = \mathcal{F}(U_0) \times \mathcal{F}(U_1) \times \cdots \times \mathcal{F}(U_n)$$

and

$$C^1(\mathcal{U}, \mathcal{F}) = \prod_{0 \le i < j \le n} \mathcal{F}(U_i \cap U_j).$$

Now we define the *coboundary map*. It is a map

$$d^p : \; C^p(\mathcal{U}, \mathcal{F}) \to C^{p+1}(\mathcal{U}, \mathcal{F})$$

defined in the following way. If c is a p-cochain then it is given as

$$c = (c_{\sigma_1}, c_{\sigma_2}, \dots, c_{\sigma_r}), \quad c_{\sigma_k} \in \mathcal{F}(U_{\sigma_k}).$$

For the components of $d^p(c)$, we set

$$d^p(c)_\sigma = d^p(c)_{(i_0, i_1, \dots, i_p, i_{p+1})} := \sum_{k=0}^{p+1} (-1)^k c_{(i_0, i_1, \dots, \hat{i}_k, \dots, i_{p+1})} | U_\sigma$$

Here we use $\hat{i}_k$ to denote the fact that we omit this index and hence get an index of length $p+1$.

Let us write this explicitly for small p values. If

$$c = (f_i)_i \in C^0(\mathcal{U}, \mathcal{F}), \quad f_i \in \mathcal{F}(U_i)$$

then

$$d^0(c) = (f_{ij})_{i<j} \in C^1(\mathcal{U}, \mathcal{F}), \quad f_{ij} \in \mathcal{F}(U_i \cap U_j)$$

with

$$f_{ij} = f_{j,|} - f_{i,|} \ .$$

For

$$c = (f_{ij})_{i<j} \in C^1(\mathcal{U}, \mathcal{F}), \quad f_{ij} \in \mathcal{F}(U_i \cap U_j)$$

we get

$$d^1(c) = (f_{ijk})_{i<j<k} \in C^2(\mathcal{U}, \mathcal{F}), \quad f_{ijk} \in \mathcal{F}(U_i \cap U_j \cap U_k)$$

with

$$f_{ijk} = f_{jk,|} - f_{ik,|} + f_{ij,|} \ .$$

If we compose two of the coboundary maps we get

$$d^{p+1} \circ d^p = 0 \ ,$$

which is easily calculated. Let us verify this in the above example:

$$\begin{aligned}
d^1(d^0((f_i)_i)) &= d^1((f_j - f_i)_{i<j}) \\
&= ((f_k - f_j) - (f_k - f_i) + (f_j - f_i))_{i<j<k} \\
&= (0)_{i<j<k}
\end{aligned}$$

We define the (Čech) p-cocycles as the kernel of d^p, the subgroup of p-cochains which are mapped to zero, and the p-coboundaries as the subgroup of the elements which are images under d^{p-1}. Because $d^p \circ d^{p-1} = 0$ we know that the coboundaries are cocycles, hence we are able to define the (Čech) cohomology groups as

$$H^p(\mathcal{U}, \mathcal{F}) := \frac{\ker d^p}{\operatorname{im} d^{p-1}}.$$

(We set $d^{-1} := 0$).

Let us calculate

$$H^0(\mathcal{U}, \mathcal{F}).$$

Because $d^{-1} = 0$ we have only to determine the elements $(f_i)_i$ with

$$(f_j - f_i)_{|U_i \cap U_j} = 0 \quad \text{for } i < j \ .$$

But for such a collection $f_i \in \mathcal{F}(U_i)$ we can find by definition of the sheaf $\mathcal{F}$ a $f \in \mathcal{F}(M)$ with $f_{|U_i} = f_i$. Conversely, starting with such a $f \in \mathcal{F}(M)$ we get by restricting it to each U_i a 0-cocycle. Hence we have

$$H^0(\mathcal{U}, \mathcal{F}) \cong \mathcal{F}(M).$$

Now we have

Theorem 8.4. *Let M be an algebraic variety, $\mathcal{U}$ an open affine covering, $\mathcal{F}$ a coherent sheaf on M, then the cohomology groups*

$$H^p(M, \mathcal{F})$$

defined by derived functors are equal to the Čech cohomology groups defined by the covering $\mathcal{U}$.

This shows in addition that the Čech cohomology groups are in fact independent on the covering chosen. For this reason we drop explicit mention of the covering. In the case of analytic or differentiable sheaves or even in the case of sheaves of abelian groups there are similar *comparison theorems.*[4]

Let us note some results.

(1) Let $\mathcal{F}$ be a coherent sheaf on a projective variety of dimension n (complex dimension), then $H^k(M, \mathcal{F})$ is a finite-dimensional vector space and

$$H^k(M, \mathcal{F}) = \{0\}, \qquad \text{for} \quad k > n. \tag{8.3}$$

(2) For an affine variety M and a coherent sheaf $\mathcal{F}$ we have

$$H^k(M, \mathcal{F}) = 0, \quad \text{if } k \geq 1.$$

(3) If M admits a finite affine covering then there exists a number k, such that for all sheaves $\mathcal{F}$ of abelian groups

$$H^p(M, \mathcal{F}) = 0, \quad \text{if } p \geq k.$$

(4) If we have a short exact sequence of sheaves

$$0 \to \mathcal{F} \to \mathcal{G} \to \mathcal{H} \to 0,$$

then there exists a long exact sequence of abelian groups (the *cohomology sequence*)

$$
\begin{aligned}
0 \to \quad & \mathcal{F}(M) \quad \to \quad \mathcal{G}(M) \quad \to \quad \mathcal{H}(M) \\
\to \; & H^1(M, \mathcal{F}) \to H^1(M, \mathcal{G}) \to H^1(M, \mathcal{H}) \\
\to \; & H^2(M, \mathcal{F}) \to \quad \cdots .
\end{aligned}
$$

Let us apply this to our "exponential sequence" (8.1)

$$0 \longrightarrow \mathbb{Z} \xrightarrow{\; i \;} \mathcal{O}^{an} \xrightarrow{\; e \;} \mathcal{O}^{*,an} \longrightarrow 0$$

in the case of a compact Riemann surface M. The cohomology of the sheaf $\mathbb{Z}$ is the usual topological cohomology. We obtain

$$
\begin{aligned}
0 \to \; & H^0(M, \mathbb{Z}) \to H^0(M, \mathcal{O}^{an}) \to H^0(M, \mathcal{O}^{*,an}) \\
\to \; & H^1(M, \mathbb{Z}) \to H^1(M, \mathcal{O}^{an}) \to H^1(M, \mathcal{O}^{*,an}) \\
\to \; & H^2(M, \mathbb{Z}) \to \quad 0 .
\end{aligned}
$$

---

[4] Forster, [Fo], S. 93, Leray's theorem.

If we calculate the known terms we get

$$
\begin{array}{ccccc}
0 \to & \mathbb{Z} & \to & \mathbb{C} & \to & \mathbb{C} \setminus \{0\} \\
\to & H^1(M,\mathbb{Z}) & \to & H^1(M,\mathcal{O}^{an}) & \to & H^1(M,\mathcal{O}^{*,an}) \\
\to & \mathbb{Z} & \to & 0 &
\end{array}
\tag{8.4}
$$

The results in the first line are consequences of the fact that M is compact and connected. In particular the only globally defined holomorphic functions are the constants. This shows that the first line is indeed exact. (Attention: this is not true anymore for noncompact Riemann surfaces.) Also $H^2(M,\mathcal{O}^{an}) = 0$ due to (8.3). It remains the exact sequence

$$
0 \to H^1(M,\mathbb{Z}) \to H^1(M,\mathcal{O}^{an}) \to H^1(M,\mathcal{O}^{*,an}) \to \mathbb{Z} .
\tag{8.5}
$$

In Sect. 8.1 we saw that vector bundles are determined by cocycle classes of matrices C_{ji}. In the case of line bundles these are just a collection of regular functions obeying the cocycle conditions. One of the condition says that they are invertible, which implies that they have no zeros in U_{ij}. If we choose an affine covering of our variety which is at the same time trivializing for the bundle, then this cocycle is a Čech 1-cocycle of the sheaf $\mathcal{O}^*$. The cohomology class of the cocycle is obtained by dividing out the 0-coboundary, hence

$$
\{\text{isomorphy classes of line bundles}\} \cong H^1(M,\mathcal{O}^*).
$$

This is also true in the analytic context. Much more could be said in this direction. Let me only remark that in this case the above map from the cohomology sequence (8.5)

$$
H^1(M,\mathcal{O}^{*,an}) \to \mathbb{Z}
\tag{8.6}
$$

gives the Chern number of the line bundle (resp. its degree). The kernel of this map corresponds to line bundles of degree 0. In particular it is isomorphic to $\text{Pic}^0(M)$. Strictly speaking, this should be the analytic Picard group. But in view of Theorem 6.4 (Chow) there is no need to distinguish between analytic or algebraic line bundles as the isomorphy classes of analytic line bundles can be identified with the isomorphy classes of algebraic line bundles. The same identification is true for the cohomology spaces of the line bundles. From (8.5) it follows that $\text{Pic}^0(M)$ can be identified with the image of $H^1(M,\mathcal{O}^{an})$ under the map e on the cohomology induced by the exponential map. If we go one step further to the left we obtain

$$
\text{Pic}^0(M) = e(H^1(M,\mathcal{O}^{an})) \cong H^1(M,\mathcal{O}^{an})/H^1(M,\mathbb{Z}) .
\tag{8.7}
$$

Note that for compact Riemann surfaces we have

$$
H^1(M,\mathbb{Z}) = (H_1(M,\mathbb{Z}))^* .
\tag{8.8}
$$

Let K be the canonical line bundle. Its sheaf of sections is the sheaf of differentials. In particular for the space of global sections we obtain $H^0(M,K) = \Omega(M)$ in the terminology of Sect. 4.1. Using Serre duality (8.11) we get

$$\mathrm{H}^1(M, \mathcal{O}^{an}) \cong (\mathrm{H}^0(M, K))^* = \mathrm{Hom}(\Omega(M), \mathbb{C}) \,. \tag{8.9}$$

Hence, we obtain in accordance with (5.2) again

$$\mathrm{Pic}^0(M) \cong \mathrm{Jac}\,(M) \cong \mathrm{H}^0(M, K)^* / \mathrm{H}_1(M, Z) \,. \tag{8.10}$$

For the following let M be an arbitrary projective variety.

(1) If M is smooth of dimension n then we can construct ω, the sheaf of n-differential forms. It is also called the canonical sheaf. It is the sheaf of sections of the determinant of the cotangent bundle. *Serre duality* says

$$\mathrm{H}^k(M, \mathcal{F}^* \otimes \omega)^* \cong \mathrm{H}^{n-k}(M, \mathcal{F}) \tag{8.11}$$

if $\mathcal{F}$ is a locally free sheaf.

(2) If L is a line bundle on M admitting a set of global sections $s_1, s_2, \ldots, s_n$ which do not vanish at a common point then we can define with them a map from M to $\mathbb{P}^{n-1}$ in the following way. Let U be a trivializing open set. Then the s_k are represented by regular functions f_k on U. We set

$$M \to \mathbb{P}^{n-1}, \quad x \mapsto (f_1(x) : f_2(x) : \ldots : f_n(x)).$$

If we change the trivialization of course the f_k will change. But fortunately all of the $f_k(x)$ are multiplied by the common number $C_{ji}(x)$ (if C_{ji} is the defining cocycle for bundle L), hence they define the same point in projective space. If a line bundle admits a set of sections in such a way that this map is an imbedding then we call L a *very ample* line bundle. This gives some new interpretation of the embedding of abelian varieties into projective space (see Chap. 6) via theta functions. Theta functions can be considered as a basis of the sections of a suitable theta bundle on the abelian variety.

(3) In the case of line bundles on smooth projective varieties we have another interesting correspondence:
The isomorphy classes of line bundles are in 1:1 correspondence to the linear equivalence classes of divisors. We will study this relation in greater detail in Chap. 9. In this case the canonical sheaf ω is locally free, hence a line bundle, called the canonical line bundle. . The corresponding divisor class is called the canonical divisor class K. As it is common use I will freely switch between the different notations ω and K.

Hints for Further Reading

Of course there are a lot of books on the subject.

[We]   Wells, R.O., *Differential Analysis on Complex Manifolds*, Springer, 1980 gives a very good introduction in Chapters I and II (pp. 1–64). He uses the differentiable and complex analytic language. Alas he does not introduce Čech cohomology. This you can find in the book already mentioned:

[Fo]   Forster, O., *Riemannsche Flächen*, Springer, 1977

or its English translation

[Foe] Forster, O., *Lectures on Riemann Surfaces*, Springer, 1981.

For a purely algebraic geometric language you should consult II.1 and III.1–4 of

[Ha]   Hartshorne, R., *Algebraic Geometry*, Springer, 1977

or

[Ii]   Iitaka, S., *Algebraic Geometry*, Springer, 1981.

A standard reference is also

[God] Godement, R., *Theorie des faisceaux*, Hermann, Paris, 1964.

9

The Theorem of Riemann–Roch for Line Bundles

9.1 Divisors and Line Bundles

In this paragraph we want to discuss the natural equivalence between classes of divisors (modulo linear equivalence) and isomorphy classes of algebraic (holomorphic) line bundles on a nonsingular projective variety. Here (as usual) we make no distinction between the line bundles and the locally free sheaves of rank 1. With this equivalence we can now reinterpret the theorem of Riemann–Roch which we formulated in Chap. 3 in terms of divisors and functions of line bundles and their sections.

We will restrict ourselves to a Riemann surface X (which is assumed to be compact as usual) and a holomorphic line bundle L on X. First we have to introduce what we mean by a *meromorphic section* of the bundle L. For this let $\{U_i\}_{i=1,\ldots,n}$ be a trivializing covering of X for the bundle L. If s is a holomorphic section of L over the open set U, then

$$s_{|U_i \cap U}$$

is given via the trivialization map by holomorphic functions f_i on $U_i \cap U$ for every set U_i. Now let A be a finite set of points of X and let s be a section over $U \setminus A$. We call s a meromorphic section over U if $s_{|U_i \cap (U \setminus A)}$ is given by meromorphic functions $f_i \in \mathcal{M}(U_i)$ for every U_i. For example, the meromorphic differentials defined in Chap. 4 are meromorphic sections of the canonical bundle.

Let s be a meromorphic section over X which is not the zero section,

$$\{f_i\}_{i=1,\ldots,n}, \qquad f_i \in \mathcal{M}(U_i)$$

be its representation by meromorphic functions. If $p \in X$ is a point in U_i then we define the order of s at the point p by

$$\mathrm{ord}_p(s) := \mathrm{ord}_p(f_i).$$

M. Schlichenmaier, The Theorem of Riemann–Roch for Line Bundles. In: M. Schlichenmaier, An Introduction to Riemann Surfaces, Algebraic Curves and Moduli Spaces, Theoretical and Mathematical Physics, 103–132 (2007)
DOI 10.1007/978-3-540-71175-9_10

Of course we have to show that this is well defined. Let p be a point lying in U_i and U_j, and let the section s be represented by f_i and f_j if restricted to $U_i \cap U_j$. Then we get

$$f_i = c_{ij} f_j.$$

Here c_{ij} is the cocycle matrix for the bundle L, which we will refer to as the transition function because it is just a 1×1 matrix. The *transition function* is a nowhere vanishing holomorphic function on $U_i \cap U_j$, hence

$$\mathrm{ord}_p(f_i) = \mathrm{ord}_p(f_j).$$

If we choose another trivialization, then the representative f_i is just multiplied by a nowhere vanishing holomorphic function in every set U_i. Hence the definition is independent of the trivialization. We can assign to every meromorphic section s of L a *divisor* by

$$(s) := \sum_{p \in X} \mathrm{ord}_p(s) \cdot p.$$

Here again only finitely many points occur in the sum. We use the following:

Proposition 9.1.[1] *Every holomorphic line bundle has a meromorphic section which is not the zero section.*

Hence we can assign to every bundle at least one divisor. Let us take two sections s_1 and s_2 of the bundle L. In general (s_1) and (s_2) will differ. But

Proposition 9.2. *The divisors (s_1) and (s_2) will belong to the same linear equivalence class. This says there will be a meromorphic function h such that*

$$(s_1) = (s_2) + (h).$$

Proof. Let f_i and g_i be representatives of the sections s_1 and s_2 by meromorphic functions on U_i. Now

$$h_i = \frac{f_i}{g_i}, \qquad h_j = \frac{f_j}{g_j}$$

are meromorphic functions on U_i, resp. U_j. But

$$f_j = c_{ji} f_i \quad \text{and} \quad g_j = c_{ji} g_i$$

hence

$$h_j = \frac{c_{ji} f_i}{c_{ji} g_i} = \frac{f_i}{g_i} = h_i, \quad \text{on} \quad U_i \cap U_j.$$

This means that the h_i glue together to form a globally defined meromorphic function h on X. If $p \in U_i$ then

$$\mathrm{ord}_p(h) = \mathrm{ord}_p(h_i) = \mathrm{ord}_p(f_i) - \mathrm{ord}_p(g_i) = \mathrm{ord}_p(s_1) - \mathrm{ord}_p(s_2),$$

which is the claim. □

---

[1] For the proof see Forster, [Fo], S. 201.

In this way we have a well-defined map, denoted by c_1, assigning to every line bundle a divisor class by taking the linear equivalence class of an arbitrary section which is not the zero section. Taking an isomorphic bundle does not change this assignment, because an isomorphy of bundles can always be described by a change of trivialization and for this we have already seen the independence.

Now the set of isomorphy classes of line bundles is an abelian group if we take as group operation the tensor product. The inverse element to L is given by its dual L^*. Remember, in Chap. 8 we saw that the tensor product of two line bundles is obtained by multiplying the transition functions involved, the dual is obtained by inverting the transition function. On the other hand the divisor classes also form an abelian group, by adding the divisors.

Proposition 9.3. *The map c_1 is an injective group homomorphism of abelian groups.*

Proof. Let L and M be line bundles, s, resp. t, be meromorphic sections. We take a common trivialization $\{U_i\}_{i=1,\dots,n}$ and represent the sections by functions f_i, resp. g_i. Then $s \otimes t$ is a section of the bundle $L \otimes M$ and it is represented by the product of functions $f_i \cdot g_i$ over U_i. We calculate the order at $p \in U_i$:

$$\mathrm{ord}_p(s \otimes t) = \mathrm{ord}_p(f_i \cdot g_i) = \mathrm{ord}_p(f_i) + \mathrm{ord}_p(g_i) = \mathrm{ord}_p(s) + \mathrm{ord}_p(t).$$

Hence

$$(s \otimes e) = (s) + (t)$$

or by considering the classes

$$c_1(L \otimes M) = c_1(L) + c_1(M).$$

Now let (s) be a principal divisor which is the divisor of a section of a bundle L. If we are able to show that L has to be isomorphic to the trivial line bundle then the map c_1 will be injective. Because (s) is a principal divisor there is a meromorphic function h such that $(s) = (h)$. Meromorphic sections of line bundles can be multiplied by meromorphic functions. The result is again a meromorphic section of the same bundle. We consider

$$r := \frac{1}{h} \cdot s$$

and calculate

$$(r) = (s) + \left(\frac{1}{h}\right) = (s) - (h) = 0.$$

Hence the section r has neither zeros nor poles. Let r be represented by functions k_i. The functions k_i are nowhere vanishing holomorphic functions on U_i. Hence we can change the trivialization map (multiply by $1/k_i$) such that the new representative is the function 1. This is possible for all U_i. Hence the line bundle can also be defined by the cocycle $c_{ij} = 1$ and hence it has to be the trivial bundle. This was the claim. □

Of course
$$c_1(L^*) = -c_1(L). \tag{9.1}$$

In fact we will show that c_1 is an isomorphism. For this it is enough to show that there is for every point p in X a line bundle L_p with

$$c_1(L_p) = [p].$$

(We use $[D]$ to denote the class of the divisor D.) Taking this for granted, for every divisor

$$D = \sum_{p \in X} n_p \cdot p$$

we can take the bundle

$$M = \bigotimes_{p \in X} L_p^{\otimes n_p},$$

where $L_p^{\otimes n_p}$ means $(L_p^*)^{\otimes |n_p|}$ if n_p is negative. Now

$$c_1(M) = c_1\left(\bigotimes_{p \in X} L_p^{\otimes n_p}\right) = \sum_{p \in X} n_p c_1(L_p) = \sum_{p \in X} n_p \cdot [p] = [D].$$

This shows the surjectivity of c_1.

For constructing such L_p we take standard coordinate patches $\{U_i\}_{i=0,\ldots,n}$ in such a way that $p \in U_0$, $p \notin U_i$, for $i = 1, \ldots, n$ and z is the coordinate in U_0 with $z(p) = 0$.

We define the line bundle by the cocycle

$$c_{i0} = \frac{1}{z}, \quad c_{0i} = z, \quad \text{for } i = 1, \ldots, n$$
$$c_{00} = 1, \quad c_{ij} = 1, \quad i, j = 1, \ldots, n .$$

Because $p \notin U_i \cap U_0$ for $i \neq 0$ these are holomorphic functions without zeros on the relevant intersections. It is easily verified by direct calculation that they respect the cocycle conditions . Hence they define a line bundle. A section s of this bundle is given by the local functions

$$f_0 = z, \quad \text{and} \quad f_i = 1, \; i = 1, \ldots n.$$

(Again this is verified by direct computations. For example, $z = z \cdot 1$, hence $f_0 = c_{0i} f_i$.)

Now
$$\mathrm{ord}_p(s) = 1, \quad \mathrm{ord}_q(s) = 0, \; q \neq p.$$

Hence
$$(s) = [p]$$

which had to be shown.

What we have shown above is valid on every nonsingular projective variety.

Theorem 9.4.[2] *Let X be a smooth projective variety. The group of isomorphy classes of line bundles over X is isomorphic to the group of linear equivalence classes of (Weil-)divisor, via the map c_1 which assigns to a section s of the line bundle its divisor class.*

Let me remark that the above theorem is not necessarily valid for singular varieties and for compact, complex manifolds which are not projective.

Let us return to our Riemann surface X. If K is the canonical divisor class of X (see Chap. 4), then a divisor is an element of K if and only if it is a divisor of a meromorphic differential. But the meromorphic differentials are sections of the holomorphic cotangent bundle ω. Hence

$$c_1(\omega) = K.$$

ω is also called the *canonical bundle* of X.

Now we want to reformulate the theorem of Riemann–Roch. Let L be a line bundle and $c_1(L) = [D]$ the associated divisor class, where

$$D = (s) = \sum_{p \in X} n_p \cdot p$$

is the divisor of a fixed meromorphic section s of L. In Chap. 3 we defined for the divisor D the following vector space of meromorphic functions:

$$L(D) := \{f \in \mathcal{M}(X) \mid (f) \geq -D\}.$$

$H^0(X, L)$ was defined in Chap. 8 as the vector space of holomorphic sections of the bundle L. Now

Proposition 9.5.

$$L(D) \cong H^0(X, L)$$

and the isomorphism is given by $f \mapsto f \cdot s$.

Proof. Take $f \in L(D)$, then $f \cdot s$ is again a meromorphic section of L. Now

$$\mathrm{ord}_p(f \cdot s) = \mathrm{ord}_p(f) + \mathrm{ord}_p(s) \geq -n_p + n_p = 0$$

for all points p. Hence $f \cdot s$ is a holomorphic section. Conversely if t is a holomorphic section then

$$f = \frac{t}{s}$$

is a meromorphic function (again: take representatives, divide them and glue together). We calculate

$$\mathrm{ord}_p(f) = \mathrm{ord}_p(t) - \mathrm{ord}_p(s) = \mathrm{ord}_p(t) - n_p \geq 0 - n_p = -n_p.$$

Hence $f \in L(D)$. Both maps are obviously inverse to each other. □

---

[2] See Hartshorne, [Ha], Cor. II, 6.16.

We define the *degree of the line bundle* L as the degree of the associated divisor class

$$\deg L = \deg(c_1(L)) = \sum_{p \in X} n_p.$$

Of course, we can take a meromorphic section s and define the degree of L directly as

$$\#(\text{of zeros of } s) - \#(\text{of poles of } s)$$

counted with multiplicities. Sometimes this is also called the *Chern number* of the bundle. It can be defined in many different ways. One is by the cohomology sequence of the exponential sequence (see (8.6)). Another one is by introducing a connection in the bundle. This connection fixes a Chern form $cf(L)$, which is a (1,1)-differential form. And then[3]

$$\int_X cf(L) = \deg L.$$

Now Riemann–Roch says (g is the genus of the Riemann surface)

$$\dim L([D]) - \dim L(K - [D]) = \deg[D] - g + 1.$$

We apply this for $D = (s)$ and by using $\deg[D] = \deg L$ and

$$c_1(\omega \otimes L^*)) = c_1(\omega) - c_1(L) = K - [D]$$

and Proposition 9.5, we obtain

Theorem 9.6. *(Riemann–Roch theorem for line bundles)*

$$\dim H^0(X, L) - \dim H^0(X, \omega \otimes L^*) = \deg L - g + 1 , \qquad (9.2)$$
$$\dim H^0(X, L) - \dim H^1(X, L) = \deg L - g + 1 . \qquad (9.3)$$

To deduce (9.3) from (9.2) we use Serre duality (8.11):

$$H^1(X, L) \cong H^0(X, \omega \otimes L^*)^*.$$

We can translate our results on the existence of meromorphic functions to the existence of holomorphic sections of certain line bundles.
Let me just formulate one result:

$$\dim H^0(X, L) \begin{cases} = 0, & \deg L < 0 \\ \geq 1 - g + \deg L, & 0 \leq \deg L < 2g - 1 \qquad (9.4) \\ = 1 - g + \deg L, & \deg L \geq 2g - 1. \end{cases}$$

If we have a line bundle L we can also ask what is the dimension of the vector space of meromorphic sections with bounded pole order or minimal

---

[3] See Wells, [We], p. 97.

zero order at finitely many fixed points. Riemann–Roch again answers this. For example let p_i, $i = 1, \ldots, k$, be points and n_i, $i = 1, \ldots, k$, be integers. We set

$$R = \bigotimes_i L_{p_i}^{\otimes n_i},$$

where L_{p_i} is the line bundle constructed above which has a section with exactly one zero at the point p_i and vanishes nowhere else. Let us denote this section by s_{p_i}. We set

$$M = L \otimes R, \quad \text{resp.} \quad L = M \otimes R^*.$$

If s is a section of M then we can multiply it in the following way:

$$t = \frac{s}{\prod_i s_{p_i}^{n_i}}.$$

t is a section of L. We calculate

$$\mathrm{ord}_q(t) = \mathrm{ord}_q(s) - \left(\sum_i n_i \, \mathrm{ord}_q(s_{p_i}) \right) = \mathrm{ord}_q(s), \quad q \neq p_i, \ i = 1, \ldots, k$$

$$\mathrm{ord}_{p_j}(t) = \mathrm{ord}_{p_j}(s) - \left(\sum_i n_i \, \mathrm{ord}_{p_j}(s_{p_i}) \right) = \mathrm{ord}_{p_j}(s) - n_j.$$

By this multiplication the space $\mathrm{H}^0(X, M)$ of holomorphic sections of M is isomorphic to the space of meromorphic sections of the bundle L which are holomorphic outside the points p_i and have at most a pole of order n_i at the points p_i, $i = 1, \ldots, n$ (see above calculation). Of course we can also choose the n_i as negative integers without disturbing the argument. In this case the section of L has a zero at the point p_i of order at least n_i.

9.2 An Application: The Krichever–Novikov Algebra

As an example how to use the machinery of Riemann–Roch for line bundles we will take the paper of Krichever and Novikov "Algebras of Virasoro Type, Riemann Surfaces and Structure of the Theory of Solitons" [KN] and show some of the basic facts which were used in this paper. We do not attempt to give a complete treatment of the content of the paper.

One of the objects of interest in string theory is the set of vector fields on a Riemann surface. If we want to use holomorphic vector fields we see that there are not many around in the case of higher genus.

But let us start with $X = \mathbb{P}^1$, the genus 0 case. Holomorphic vector fields are sections of the (holomorphic) tangent line bundle T. This bundle is dual to the cotangent bundle ω, which itself corresponds to the canonical divisor class K. Hence

$$\deg T = \deg(-K) = -(2g - 2) = 2 - 2g.$$

In the case of genus 0 we get

$$\deg T = 2 \geq 2g(\mathbb{P}^1) - 1.$$

Riemann–Roch says

$$\dim \mathrm{H}^0(X,T) = \deg T - g(\mathbb{P}^1) + 1 = 3.$$

Hence the vector space of holomorphic vector field has the dimension 3. To calculate a basis we use the standard covering of $\mathbb{P}^1$ already introduced in Sect. 1.2:

$$U_0 := \mathbb{P}^1 \setminus \{\infty\} \cong \mathbb{C}, \quad U_1 := \mathbb{P}^1 \setminus \{0\} \cong \mathbb{C}.$$

The coordinates are z in U_0 and w in U_1. The transformation is given by

$$w = \frac{1}{z}.$$

We saw in Sect. 4.1 that dz is a globally defined meromorphic differential which is represented by the pair of meromorphic functions

$$\left(1, \frac{-1}{w^2}\right).$$

The pair of the inverted functions now represents a (meromorphic) vector field

$$(1, -w^2).$$

Because the functions have no poles it is in fact a holomorphic vector field. Now it is easy to give by multiplication two more linearly independent holomorphic vector fields:

$$(z, -w) \quad \text{and} \quad (z^2, -1).$$

In their local description on U_1 they can also be given as follows:

$$\frac{\partial}{\partial z}, \quad z\frac{\partial}{\partial z}, \quad z^2\frac{\partial}{\partial z}.$$

Obviously they are a basis of $\mathrm{H}^0(\mathbb{P}^1, T)$.

The vector space of vector fields is a Lie algebra with respect to the Lie bracket

$$[X, Y] := X \circ Y - Y \circ X.$$

In our case we calculate

$$\left[z\frac{\partial}{\partial z}, \frac{\partial}{\partial z}\right] = z\frac{\partial}{\partial z}\left(\frac{\partial}{\partial z}\right) - \frac{\partial}{\partial z}\left(z\frac{\partial}{\partial z}\right) = -\frac{\partial}{\partial z},$$

$$\left[z\frac{\partial}{\partial z}, z^2\frac{\partial}{\partial z}\right] = z\frac{\partial}{\partial z}\left(z^2\frac{\partial}{\partial z}\right) - z^2\frac{\partial}{\partial z}\left(z\frac{\partial}{\partial z}\right) = z^2\frac{\partial}{\partial z},$$

$$\left[z^2\frac{\partial}{\partial z}, \frac{\partial}{\partial z}\right] = z^2\frac{\partial}{\partial z}\left(\frac{\partial}{\partial z}\right) - \frac{\partial}{\partial z}\left(z^2\frac{\partial}{\partial z}\right) = -2z\frac{\partial}{\partial z}.$$

This is a well-known Lie algebra. If we make a change of basis

$$H = 2z\frac{\partial}{\partial z}, \quad X = iz^2\frac{\partial}{\partial z}, \quad Y = i\frac{\partial}{\partial z}$$

we see

$$[H, X] = 2X, \quad [H, Y] = -2Y, \quad [X, Y] = H.$$

These are the defining relations for the Lie algebra $sl\,(2,\mathbb{C})$ of traceless complex 2×2 matrices with the commutator as Lie bracket.

As a next example we take the torus T. Now $g(T) = 1$. Moreover dz is a differential without zeros on T (we use the notation of Chap. 4). This says: the bundle ω is the trivial bundle $\mathcal{O}$. Hence the tangent bundle is also trivial. The holomorphic vector fields are just

$$c \cdot \frac{\partial}{\partial z}, \quad c \in \mathbb{C}.$$

For X a Riemann surface of genus $g \geq 2$ we have $\deg T = -2g + 2$. Hence there will be no holomorphic vector field different from zero. To get a nontrivial vector field we have to allow poles. We fix points P_+ and P_- in "general position" on X and coordinates z_+, resp. z_-, around these points, i.e.

$$z_+(P_+) = 0 \quad \text{and} \quad z_-(P_-) = 0\,.$$

What we understand by "general position" we will describe later on.

We denote by $\mathcal{L}$ the space of meromorphic vector fields which are holomorphic outside the points P_+ and P_-. Of course this vector space is closed under the Lie bracket, hence these vector fields constitute a Lie algebra. We want to exhibit a special basis

$$\{e_i\}_{i\in\mathbb{Z}}$$

of this space.[4] We restrict ourselves to the case $g \geq 2$. In the case $g = 1$ we need a slight modification of the basis (see the article for the details).

Let us take the bundles

$$M_i = T \otimes L_{P_+}^{-i-1} \otimes L_{P_-}^{i+3g-1} \quad i \in \mathbb{Z}.$$

(The bundles $L_{P_\pm}$ are the bundles corresponding to the points $P_\pm$ as they were introduced in Sect. 9.1.) We calculate

$$\deg M_i = (2 - 2g) + (-i - 1) + (i + 3g - 1)$$
$$= 2 - 2g + 3g - 2 = g.$$

---

[4] In view of the multi-point generalization we shifted the index used in the original article [KN].

Riemann–Roch says now

$$\dim \mathrm{H}^0(X, M_i) \geq \deg M_i - g + 1 = 1.$$

Hence, by reinterpretation of the sections of M_i, there exists for every $i \in \mathbb{Z}$ a vector field which is holomorphic outside the points $P_\pm$ and has at least order

$$(i + 1) \quad \text{at } P_+$$

and at least order

$$(i + 3g - 1) \quad \text{at } P_-.$$

Of course, positive orders correspond to zeros, negative orders to poles.
   For generic choices of $P_\pm$ we will show that

$$\dim \mathrm{H}^0(X, M_i) = 1$$

and the corresponding (up to a scalar multiple) unique meromorphic section of T has exactly the above multiplicities. We denote this section by e_i.
   To prove this claim we have to introduce Weierstraß points with regard to quadratic differentials (2-Weierstraß points). Let K be the canonical divisor class, then $2K$ is the associated divisor class for the quadratic differentials. Because $\deg(2K) = 4g - 4 \geq 2g - 1$ (remember we assume $g > 1$) we get by Riemann–Roch that $\dim \mathrm{H}^0(X, 2K) = 3g - 3$. Let $P \in X$ be a point. Our aim is to calculate for $n \geq 0$ $\dim \mathrm{H}^0(X, 2K - nP)$. This number gives the dimension of the space of quadratic differentials having a zero of order $\geq n$ at P. Now $\deg(2K - nP) = 4g - 4 - n$. If $n \leq 2g - 3$ the degree is $\geq 2g - 1$, hence

$$\dim \mathrm{H}^0(2K - nP) = 4g - 4 - n - g + 1 = (3g - 3) - n \; (\geq g).$$

If $n \geq 4g - 3$ then $\deg(2K - nP) < 0$, hence $\dim \mathrm{H}^0(2K - nP) = 0$.
In the region $2g - 3 < n < 4g - 3$ the dimensions of the vector spaces have to drop from g to 0.
Let D be a divisor with $\dim \mathrm{H}^0(X, D) = r$. Then we have always

$$r \leq \dim \mathrm{H}^0(X, D + P) \leq r + 1.$$

To see this let f, h be in $\mathrm{H}^0(X, D + P)$ but not in $\mathrm{H}^0(X, D)$. We can subtract the additional pole at the point P of h by subtracting a suitable multiple of f. Hence the dimension can increase only by 1.
In our case this says that the dimension will drop at most by one in every step. Let us consider the points where this sequence of dimensions will be

$$\dim \mathrm{H}^0(X, 2K - nP) = 3g - 3 - n, \quad \text{for } n \leq 3g - 3. \tag{9.5}$$

Hence necessarily we have

$$\dim \mathrm{H}^0(X, 2K - nP) = 0 \quad \text{for } n \geq 3g - 3. \tag{9.6}$$

If (9.5) is not true for the point P we call P a 2-Weierstraß point. For P to be a 2-Weierstraß point it is equivalent that there exists a quadratic differential with zero order $\geq 3g - 3$ at the point P. (The classical Weierstraß points are with regard to differentials. There the condition is: it exists a holomorphic differential with a zero of order $\geq g$ at the point P.) Fortunately there are only finitely many 2-Weierstraß points on X. We will not prove this here (see Farkas/Kra, [WE], p. 84). The essential idea is that the set of (2-)Weierstraß points can be given as the zero set of a collection of holomorphic local functions.

Hence we can always avoid these points. So let $P_\pm$ be no 2-Weierstraß points. Let us calculate $H^0(X, M_i)$. We use the additive divisor notation and drop X in the notation. We consider

$$-K + nP_+ + (3g - 2 - n)P_-$$

with $n \in \mathbb{Z}$. Because the argument is symmetric in P_+ and P_- we can restrict ourselves to $n > 0$. By Riemann–Roch we get

$$\dim H^0(-K + nP_+ + (3g - 2 - n)P_-) - \dim H^0(2K - nP_+ - (3g - 2 - n)P_-) = 1.$$

Let us consider first the case $n \geq 3g - 2$. By (9.6) we have $\dim H^0(2K - nP_+) = 0$, hence

$$\dim H^0(-K + nP_+) = n - (3g - 3),$$

in particular, $\dim H^0(-K + (3g - 2)P_+) = 1$. Let f_1 be the generator of this space. We can choose such a P_- that it is not a zero of f_1. f_1 has to have pole order $(3g - 2)$ at P_+. Otherwise f_1 would be in $H^0(-K + (3g - 3)P_+)$. But this space is zero as follows again with Riemann–Roch and (9.6). If we choose P_- general enough then

$$\dim H^0(-K + nP_+ + (3g - 2 - n)P_-) = 1,$$

and the generator has exactly the maximal pole order at both points. Let me deduce this for $n = 3g - 1$. The proof for the general case is essentially the same and goes by induction. We have $\dim H^0(-K + (3g - 1)P_+) = 2$. Let f_1 be as above and let f be a second element such that $\{f_1, f\}$ is a basis of this vector space. We can solve the equation

$$af_1(P_-) + cf(P_-) = 0$$

with $c \neq 0$. $f_2 = af_1 + cf$ is a function such that f_1, f_2 is again a basis. f_2 has at least a zero of order 1 at P_-. We do not want f_2 to have a higher order zero. For this we have to make sure that $af_1'(P_-) + cf'(P_-) \neq 0$ by choosing P_- suitable. Now f_2 generates the subspace $H^0(-K + (3g - 1)P_+ - P_-)$. It has the right zero order at P_-. If we assume that it does not have the full pole order $3g - 1$ at P_+ it would be an element of $H^0(-K + (3g - 2)P_+)$, hence it would be a multiple of f_1 in contradiction to its construction.

Let us consider the remaining cases $0 < n \leq 3g - 3$. By (9.5) we have $\dim H^0(2K - nP_+) = 3g - 3 - n$. Because P_- is not a 2-Weierstraß point we get

$$\dim H^0(2K - nP_+ - kP_-) = \begin{cases} 3g - 3 - n - k & , 0 \leq k \leq 3g - 3 - n \\ 0 & , k = 3g - 2 - n . \end{cases}$$

Now by Riemann–Roch

$$\dim H^0(-K + nP_+ + (3g - 2 - n)P_-) - 0 = 1.$$

The generator again has exactly the maximal pole order. If it has a pole order at P_+ less than the expected, it would be an element of $H^0(-K+(n-1)P_++(3g-2-n)P_-)$. But this space is zero (apply Riemann–Roch again). The argument for P_- is the same. This completes the proof.

Locally around the points $P_\pm$ the section can be given by

$$e_i = a_i^+ z_+^{i+1} (1 + O(z_+)) \cdot \frac{\partial}{\partial z_+}, \quad a_i^+ \in \mathbb{C}, \tag{9.7}$$

$$e_i = a_i^- z_-^{-i-3g+1} (1 + O(z_-)) \cdot \frac{\partial}{\partial z_-}, \quad a_i^- \in \mathbb{C}. \tag{9.8}$$

If we require $a_i^+ = 1$, then e_i is completely fixed.

We define the subspaces

$$\mathcal{L}_+ := \langle e_i \mid i \geq -1 \rangle,$$
$$\mathcal{L}_- := \langle e_i \mid i \leq -3g + 1 \rangle,$$
$$\mathcal{L}_0 := \langle e_i \mid -3g + 2 \leq i \leq -2 \rangle,$$

$\mathcal{L}_+$ is the subspace of the vector fields which are also holomorphic at the point P_+. $\mathcal{L}_-$ is the subspace of the vector fields which are also holomorphic at the point P_-.

If v is a meromorphic vector field holomorphic outside $P_\pm$ then by subtracting elements coming from $\mathcal{L}_-$ we can always obtain that the pole order at P_+ is $\leq 3g - 3$ without introducing new poles at P_-. In the same way we can subtract elements coming from $\mathcal{L}_+$ to obtain additionally that the pole order at P_- is $\leq 3g - 3$. Now we can subtract elements of $\mathcal{L}_0$ to remove the poles at P_+ and still get that the pole order at P_- is $\leq 3g - 3$. The result w is zero. If we assume w is not zero then we can set n equal to the pole order at P_- of the result (remember vector fields need to have poles). Hence w corresponds to a section of

$$M_{n-3g+1} = T \otimes L_{P_+}^{-n+3g-2} \otimes L_{P_-}^n.$$

Because it has no pole at P_+ it is not a multiple of e_{n-3g+1}. This is a contradiction to $\dim H^0(X, M_{n-3g+1}) = 1$. With this argument we showed that v is a linear combination of e_i.

Theorem 9.7. *A basis of the Lie algebra $\mathcal{L}$ of vector fields consisting of those vector fields holomorphic outside the points P_+ and P_- is given by the above e_i with $i \in \mathbb{Z}$. In this basis the Lie bracket is given by*

$$[e_i, e_j] = \sum_{k=0}^{3g} c_{ij}^k e_{i+j+k}, \quad c_{ij}^k \in \mathbb{C}, \quad c_{ij}^0 = (j - i).$$

Proof. By the above arguments $\mathcal{L}$ has as basis the $\{e_i \mid i \in \mathbb{Z}\}$. The only thing which remains to be shown is the rule for the commutator. We know the vector field given by the Lie bracket can again be developed in these e_i. To get its coefficients we have to calculate its order of the poles, resp. zeros, at the points $P_\pm$. We start with P_+. The elements e_i and e_j are given locally by (9.7). By a direct calculation we see

$$[e_i, e_j] = z_+^{(i+j+1)} \left((j - i) + O(z_+) \right) \frac{\partial}{\partial z_+}.$$

The order of the zero of this vector field is $\geq (i + j + 1)$. Hence only e_r with $r \geq (i + j)$ are involved. At the point P_- we get the result

$$[e_i, e_j] = a_i^- a_j^- z_-^{-(i+j+6g)+1} \left((i - j) + O(z_-) \right) \frac{\partial}{\partial z_-}.$$

The order of the pole is $\leq (i + j + 6g - 1)$. Hence only e_r with $r \leq (i + j + 3g)$ are involved. These two facts combined is the claim.   □

In our proof we are able to give the exact formula for the coefficients in the extremal cases:

$$c_{ij}^0 = j - i, \qquad c_{ij}^{3g} = (i - j) \frac{a_i^- a_j^-}{a_{i+j+3g}^-}.$$

This Lie algebra is called the *Krichever–Novikov (vector field) algebra*. It is not a graded algebra, but it is "almost graded". The definition of almost-gradedness is the formula in Theorem 9.7.

Obviously $\mathcal{L}_+$ and $\mathcal{L}_-$ are subalgebras. We have the decomposition

$$\mathcal{L} = \mathcal{L}_+ \oplus \mathcal{L}_0 \oplus \mathcal{L}_-.$$

$\mathcal{L}_0$ is a finite-dimensional subspace. Its dimension is $3g - 3$. In fact it can be shown that this subspace corresponds to infinitesimal analytic deformations of the Riemann surface X. Moreover it can be identified with the tangent space to the moduli space [SchSh].

In a similar manner we can construct a basis for the vector space of meromorphic sections of the bundle K^λ which are holomorphic outside the points $P_\pm$. Here K denotes the canonical bundle and λ is a multiple of $\frac{1}{2}$.

Now we would like to establish a link between these vector fields and the vector fields on the circle S^1. For this we take a meromorphic differential which has poles of order 1 at the points $P_\pm$ with residue $+1$ at P_+ and residue -1 at P_-. The existence of such a differential is guaranteed by Theorem 4.6. Of course it is not unique. But by adding a suitable global holomorphic differential we get such a differential which has only imaginary periods if we integrate it over the homology cycles.

To show that this is possible let us start with an arbitrary differential σ. Let $\alpha_i, \beta_i,\ i = 1,\ldots,g$ be a symplectic homology basis and $\omega_i,\ i = 1,\ldots,g$ the associated canonical basis of the holomorphic differentials, see Chap. 5. We have

$$\int_{\alpha_i} \omega_j = \delta_{ij}, \quad \int_{\beta_i} \omega_j = \pi_{ij}, \quad \Pi := (\pi_{ij}).$$

The matrix $\mathbf{Im}\,\Pi$ is a symmetric positive definite $g \times g$ matrix. In particular it is nonsingular. We calculate

$$\int_{\alpha_i} \sigma = a_i + \mathrm{i}\,b_i, \quad \int_{\beta_i} \sigma = c_i + \mathrm{i}\,d_i,$$

with $a_i, b_i, c_i, d_i \in \mathbb{R}$ for $i = 1,\ldots,g$. We introduce the vectors

$$a = {}^t(a_1, a_2, \ldots, a_g), \quad c = {}^t(c_1, c_2, \ldots, c_g).$$

We are able to determine a vector $f = {}^t(f_1, f_2, \ldots, f_g)$ with

$$(\mathbf{Im}\,\Pi) \cdot f = -c + (\mathbf{Re}\,\Pi) \cdot a.$$

We set

$$\rho := \sigma - \sum_{l=1}^{g}(a_l + \mathrm{i}\,f_l)\omega_l$$

and calculate

$$\int_{\alpha_j} \rho = \int_{\alpha_j} \sigma - \sum_{l=1}^{g}(a_l + \mathrm{i}\,f_l)\int_{\alpha_j}\omega_l = a_j + \mathrm{i}\,b_j - a_j - \mathrm{i}\,f_j$$

$$= \mathrm{i}(b_j - f_j)$$

$$\int_{\beta_j} \rho = \int_{\beta_j} \sigma - \sum_{l=1}^{g}(a_l + \mathrm{i}f_l)\int_{\beta_j}\omega_l = c_j + \mathrm{i}\,d_j - \sum_{l=1}^{g}(a_l + \mathrm{i}\,f_l)\pi_{jl}$$

$$= c_j - (\mathbf{Re}\,\Pi)a)_j + ((\mathbf{Im}\,\Pi)f)_j + \mathrm{i}A_j.$$

with suitable $A_j \in \mathbb{R}$. Now f has been chosen in such a way that the above real part vanishes. Hence this ρ has only imaginary periods (all periods are integer multiples of the basic periods).

Consider ρ and η were two differentials which fulfil the conditions. In this case $\gamma := \rho - \eta$ is a holomorphic differential with imaginary periods. But such a differential has to be zero due to Riemann's bilinear relations (Sect. 5.2):

$$\mathbf{Im}\left(\sum_{i=1}^{g}\left(\overline{\int_{\alpha_i}\gamma}\cdot\int_{\beta_i}\gamma\right)\right)>0.$$

Hence such a differential ρ is uniquely determined.

We fix a point $Q_0\neq P_\pm$. Now

$$r(Q)=\int_{Q_0}^{Q}\rho$$

is not a well-defined function . But due to the above normalization at least

$$u(Q):=\mathbf{Re}\ r(Q)$$

is a well-defined harmonic function.

The contours C_τ are defined as the level lines of this function

$$C_\tau:=\{Q\in X\mid u(Q)=\tau\}.$$

Due to the prescribed poles $1/z_+$, resp. $-1/z_-$, at the points $P_\pm$ we find that C_τ is a circle around P_- for $\tau\to\infty$ and C_τ is a circle around P_+ for $\tau\to-\infty$.

If we restrict $\mathcal{L}$ to C_τ we get a homomorphism

$$\mathcal{L}\to\mathcal{L}(C_\tau),$$

where the latter denotes the algebra of vector fields on the contour. Because for big $|\tau|$ these contours are circles we get a homomorphism

$$\mathcal{L}\to\mathcal{L}(S^1).$$

Krichever and Novikov showed that the image of $\mathcal{L}$ is dense in $\mathcal{L}(S^1)$.

It is possible to construct central extensions of these vector field algebras. These central extensions are higher genus analogues of the Virasoro algebra. .

Hints for Further Reading

You find the new formulation of Riemann–Roch and the connection between line bundles and divisors in the books

[Fo]   Forster, O., *Riemannsche Flächen*, Springer, 1977.

[Foe] Forster, O., *Lectures on Riemann Surfaces*, Springer, 1981.

[GH]   Griffiths, Ph., Harris, J., *Principles of Algebraic Geometry*, Wiley, New york, 1978.

[Ha]   Hartshorne, R., *Algebraic Geometry*, Springer, 1977.

In fact many of the textbooks prove Riemann–Roch for line bundles and deduce the form of Chap. 3 as a reformulation.
The Krichever–Novikov algebra is introduced in

[KN]   Krichever, I.M., Novikov, S.P., Algebras of Virasoro Type, Riemann Surfaces and Structures of the Theory of Solitons, *Funkts. Anal. Prilozh.* **21**(2) (1987) 46; Virasoro Type Algebras, Riemann Surfaces and Strings in Minkowski Space, *Funkts. Anal. Prilozh.* **21**(4) (1987) 47.

For applications in string theory see also

[Bo]   Bonora, L., Bregola, M., Cotta-Ramusino, P., Martinelli, C., Virasoro Type Algebras and BRST Operators on Riemann Surfaces, *Phys. Lett.* **B205** (1988) 53.

The multi-point version of these algebras (including differential operator algebras, current algebras and their central extensions) was introduced and studied by Schlichenmaier. It provides a global operator approach to conformal field theory, in particular to Wess–Zumino–Novikov–Witten models. As a starting reference for these directions one might take

[Schl]   Schlichenmaier, M., Central Extensions and Semi-infinite Wedge Representations of Krichever–Novikov Algebras for more than Two Points, *Lett. Math. Phys.* **20** (1990) 33–46.

[Schl2]   Schlichenmaier, M., *Local Cocycles and Central Extensions for Multi-point Algebras of Krichever–Novikov Type*, J. Reine Angew. Math. **559** (2003) 53–94.

[Schl3]   Schlichenmaier, M., *Higher Genus Affine Algebras of Krichever–Novikov Type*, Moscow Math. J. **4** (2003) 1395–1427.

[SchSh]   Schlichenmaier, M., Sheinman, O.K., *Knizhnik–Zamolodchikov Equations for Positive Genus and Krichever–Novikov Algebras*, Russian Math. Surv. **59**(4) (2004) 737–770.

You find the relevant facts on Weierstraß points for example in

[FK]   Farkas, H.M., Kra, I., *Riemann Surfaces*, Springer, 1980.

[WE]   Wells, R.O., Differential Analysis on Complex Manifields, Springer, 1980.

10

The Mumford Isomorphism
on the Moduli Space

10.1 The Mumford Isomorphism

Let $\pi : X \to B$ be a family of nonsingular algebraic curves of genus $g \geq 2$ over a nonsingular base variety B. Recall that π is always assumed to be surjective. If E is a vector bundle, resp. its locally free sheaf of sections or even an arbitrary coherent sheaf, then we can assign to it the higher direct image sheaves $R^i\pi_*E$, $i = 0, 1, \ldots$, over the base variety B in the following way. We take the assignment

$$U \mapsto H^i(\pi^{-1}(U), E_{|\pi^{-1}(U)}), \quad i = 0, 1, \ldots$$

for every open set U of B.

For $i = 0$ we get

$$U \mapsto H^0(\pi^{-1}(U), E_{|\pi^{-1}(U)}) = E(\pi^{-1}(U)),$$

hence this defines a sheaf. We denote this sheaf by $R^0\pi_*E$ or just π_*E and call it the *direct image sheaf* of E. Unfortunately this assignment is for $i > 0$ not necessarily a sheaf. But it is a presheaf. A *presheaf* is an object similar to a sheaf which does not necessarily fulfil the conditions (3) and (4) of the definition of a sheaf in Sect. 8.2. (Roughly speaking, the local data does not glue to global data.) There is a way to assign to every presheaf a sheaf in a minimal way. This sheaf is called the *associated sheaf* of the presheaf. If the presheaf is already a sheaf then the associated sheaf is the sheaf itself.[1] Now the $R^i\pi_*E$, $i = 0, 1, 2, \ldots$, are defined to be the sheaves associated to the above presheaves. They are called *higher direct image sheaves*. Note that this construction also makes sense for families of higher dimensional varieties.

Even if E is locally free, the sheaves $R^i\pi_*E$ are not necessarily locally free. But there is at least a sufficient criteria when this will be the case. And

---

[1] See Hartshorne [Ha] for details.

M. Schlichenmaier, The Mumford Isomorphism on the Moduli Space. In: M. Schlichenmaier, An Introduction to Riemann Surfaces, Algebraic Curves and Moduli Spaces, Theoretical and Mathematical Physics, 119–132 (2007)
DOI 10.1007/978-3-540-71175-9_11

this criteria will be fulfilled in the cases which are of interest to us. Take p a point of the base variety B and calculate

$$\dim \mathrm{H}^i(\pi^{-1}(p), E_{|\pi^{-1}(p)}) = k(p).$$

If $k(p)$ is independent of the point p chosen then $\mathrm{R}^i\pi_* E$ is locally free of rank $k = k(p)$. Moreover the fibre of the vector bundle $\mathrm{R}^i\pi_* E$ at the point p can be identified in a natural way with

$$\mathrm{H}^i(\pi^{-1}(p), E_{|\pi^{-1}(p)}).$$

In our case the dimension of the fibres of the family X is equal to 1, hence

$$\mathrm{H}^i(\pi^{-1}(p), E_{|\pi^{-1}(p)}) = 0, \quad i > 1$$

and hence $\mathrm{R}^i\pi_* E = 0$ for $i > 1$.

Before we calculate examples let me introduce one of the most important locally free sheaves which exists for every family of curves. It is the sheaf of relative differentials ω_π. It is defined in the following way. Let $\pi : X \to B$ be a family of curves, X and B nonsingular, and T_X and T_B the tangent bundles. We can pullback the bundle T_B to X and get a surjective map (the map induced by the differential of π):

$$\psi : T_X \to \pi^* T_B.$$

This can seen using the interpretation of the tangent space as the space of derivations on the functions. If f is a function on B then $\pi^* f$

$$\pi^* f(x) := f(\pi(x))$$

is a function on X. If D is a derivation on X then ψD

$$(\psi D)(f) := D(\pi^* f)$$

is a derivation on B. A local calculation shows that ψ is surjective. We define the *relative tangent sheaf*

$$T_\pi := \ker \psi.$$

Because $\dim X = \dim B + 1$, it is a line bundle. Essentially its sections are the derivations which operate nontrivially only in fibre direction. If we take C a fibre (hence a curve) we get

$$T_{\pi|C} \cong T_C.$$

The sheaf of *relative differentials* ω_π is now the dual sheaf of T_π. We get

$$\omega_{\pi|C} \cong \omega_C.$$

Because we will work with a fixed family we will drop the index and just write ω.

As a first example of direct image sheaves we consider as line bundle L on X the locally free sheaf ω of relative differentials. Let $p \in B$ be a point and $C = \pi^{-1}(p)$ be the genus g curve which is the fibre above p. Now $\omega_{|C} \cong \omega_C$ and hence

$$H^0(C, \omega_{|C}) = H^0(C, \omega_C)$$
$$H^1(C, \omega_{|C}) = H^1(C, \omega_C) \cong H^0(C, \mathcal{O})^* = \mathbb{C}.$$

We used Serre duality (8.11) in the second line. Because $\dim H^0(C, \omega_C) = g$, we see that $\pi_* \omega$ is a vector bundle of rank g and

$$R^1 \pi_* \omega \cong \mathcal{O}_B \tag{10.1}$$

is the trivial line bundle on the base variety B. Here we used the fact that Serre duality is a "natural isomorphy", which says in our situation that it is valid not only at the point but universally along the base variety B.[2] Only with this additional remark we can conclude from the second row that the bundles involved are isomorphic.

As a second example we take as L the trivial bundle $\mathcal{O}_X$:

$$H^0(C, \mathcal{O}_{X|C}) = H^0(C, \mathcal{O}_C) = \mathbb{C}$$
$$H^1(C, \mathcal{O}_{X|C}) = H^1(C, \mathcal{O}_C) \cong H^0(C, \omega_C)^*.$$

Again all dimensions are invariant and we see

$$\pi_* \mathcal{O}_X = \mathcal{O}_B, \quad R^1 \pi_* \mathcal{O}_X = (\pi_* \omega)^*.$$

As a third example we consider the higher powers ω^n, $n \geq 2$ of the bundle ω:

$$H^0(C, \omega_{|C}^n) = H^0(C, \omega_C^n)$$
$$H^1(C, \omega_{|C}^n) = H^1(C, \omega_C^n) \cong H^0(C, \omega_C^{1-n})^*.$$

Now

$$\deg \omega_C^{1-n} = (1 - n)(2g - 2) < 0$$

and hence by Riemann–Roch $\dim H^0(C, \omega_C^{1-n}) = 0$. This says

$$R^1 \pi_* \omega^n = 0. \tag{10.2}$$

Riemann–Roch also says

$$\dim H^0(C, \omega_C^n) = \deg \omega_C^n - g + 1 = n(2g - 2) - g + 1$$
$$= (2n - 1)g - (2n - 1) =: k(n).$$

In particular, $\pi_*(\omega^2)$ is a bundle of rank $3g - 3$ generated at the point $p \in B$ by the global quadratic differentials on the curve C lying above p.

---

[2] See Hartshorne [Ha].

Now we define the following line bundles on B:

$$\lambda_1 = \wedge^g \pi_* \omega,$$
$$\lambda_n = \wedge^{k(n)} \pi_*(\omega^n), \ n \geq 2.$$

In fact all λ_n are tensor powers of the bundle λ_1 as we have

Theorem 10.1. *(The Mumford isomorphism)*

$$\lambda_n \cong \lambda_1^{6n^2 - 6n + 1}, \quad especially \quad \lambda_2 \cong \lambda_1^{13}.$$

We will give a proof of this in Sect. 10.2 by using the Grothendieck–Riemann–Roch theorem, which is of interest by itself. In this section we will apply the Mumford isomorphism to the moduli space as the base variety. Unfortunately neither $\mathcal{M}_g$ nor $\overline{\mathcal{M}_g}$ are nonsingular nor does there exist a universal family of curves over them, as has already been noted several times in the course of these lectures. But by using the Picard group of the moduli functor (see Chap. 7) and the desingularization of $\overline{\mathcal{M}_g}$ one can show[3] that the results are valid there. The result can again be reinterpreted in terms of the moduli space itself. So let us pretend for the following that $\mathcal{M}_g$ is nonsingular and that there is a universal family of curves

$$\pi : \mathcal{C} \to \mathcal{M}_g$$

over the moduli space. In this case λ_1 is the Hodge line bundle λ introduced in Chap. 7. What is $\pi_*(\omega^2)$? We have already noted that the tangent space at a point p in $\mathcal{M}_g$ is given in a natural way by

$$\mathrm{H}^1(C, T_C)$$

where $C = \pi^{-1}(p)$. If T is the relative tangent bundle of the family, then we saw

$$T_{|C} = T_C.$$

Hence

$$\mathrm{R}^1 \pi_* T \cong T_{\mathcal{M}_g},$$

where $T_{\mathcal{M}_g}$ denotes the tangent bundle of the moduli space (at least at the nonsingular points). But on the other hand, by Serre duality (8.11)

$$\mathrm{H}^1(C, T_{|C}) = \mathrm{H}^1(C, T_C) = \mathrm{H}^0(C, \omega_C^2)^* = \mathrm{H}^0(C, \omega^2_{|C})^*$$

and hence

$$\mathrm{R}^1 \pi_* T \cong (\pi_*(\omega^2))^*.$$

In other words $\pi_* \omega^2$ is the cotangent bundle of the moduli space. Hence the quadratic differentials of the curve C are the local differentials of the moduli space.

---

[3] See Mumford, [M], Thm. 5.10.

Now

$$\lambda_2 = \wedge^{3g-3}\pi_*(\omega^2)$$

has as section the top dimensional holomorphic differentials. This line bundle is called the *canonical bundle* $\mathcal{K}_{\mathcal{M}_g}$ (again at least at the nonsingular points). The Mumford isomorphism (we use the additive divisor notation) says in our case

Proposition 10.2.

$$\mathcal{K}_{\mathcal{M}_g} \cong 13\lambda.$$

If one makes the analysis with $\overline{\mathcal{M}}_g$ one gets the following result[4] (again at the nonsingular points)

Proposition 10.3.

$$\mathcal{K}_{\overline{\mathcal{M}}_g} \cong 13\lambda - 2\delta - \delta_1.$$

Here we use the same notation as in Sect. 7.3:

$$\delta_i := [\Delta_i], \ i = 0, 2, 3, \ldots, \left[\frac{g}{2}\right],$$

$$\delta_1 := \frac{1}{2}[\Delta_1],$$

$$\delta := \sum_i \delta_i.$$

Notice that the divisors on the right-hand side of the proposition really live on $\overline{\mathcal{M}}_g$, because

$$12\lambda , \ 2\delta \quad \text{and} \quad \lambda - \delta_1 = (\lambda + \delta_1) - 2\delta_1$$

are in fact elements of $\operatorname{Pic}\overline{\mathcal{M}}_g$ (see Proposition 7.6).

The above formulas are fundamental in the study of the Polyakov integration measure for the closed bosonic string. In spacetime dimension $D = 26$ this measure can be formulated in terms of objects on the moduli space. On $\mathcal{M}_g$ the bundle

$$\mathcal{K}_{\mathcal{M}_g} \otimes \lambda^{-13}$$

is trivial (this is the content of Proposition 10.2), hence it has a global nowhere vanishing holomorphic section μ_g. This section is unique up to multiplication with a constant. The content of the Belavin–Knizhnik theorem is the following: in the case of critical dimension $D = 26$ the integration form is equal to the square of the modulus of the section μ_g divided by the 13th power of the natural metric on the bundle λ (coming from the Hodge product on the differentials on the curve). Proposition 10.3 says how the section μ_g and the integration measure will behave if we approach the boundary of the moduli

---

[4] Harris/Mumford, [HM], p. 48.

space. For more details and the exact formulation I like to refer to the list of literature at the end of this section. Let me just quote the following from the paper of Beilinson and Manin. We assume $g \geq 2$. The Polyakov partition function of the closed bosonic string in $D = 26$ can be given, after dividing out all invariance groups, by

$$Z_g = \text{const} \cdot \int_{\mathcal{M}_g} d\nu \, \det{}' \Delta_{2\gamma} \, (\det{}' \Delta_{0\gamma})^{-13}.$$

Here γ denotes the metric of constant curvature -1 in a given conformal class. $\Delta_{2\gamma}$ and $\Delta_{0\gamma}$ are the Laplace operators on the quadratic differentials, resp. on the functions. They depend on the metric γ. The symbol $\det'$ denotes the so-called *zeta function regulated* determinant of these operators (see Bost [Bo]), and $d\nu$ is the Petersson–Weil measure on the moduli space.

Let $p \in \overline{\mathcal{M}_g}$ be a point of the moduli space and C the corresponding curve. If $w_1, w_2, \ldots, w_{3g-3}$ are a basis of the global quadratic differentials on the curve C depending holomorphically on local coordinates on $\mathcal{M}_g$ around the point p (the deformations of the curve), then they describe local 1-forms $W_1, W_2, \ldots, W_{3g-3}$ on $\mathcal{M}_g$:

$$w_1 \wedge w_2 \wedge \cdots \wedge w_{3g-3} \quad \text{resp.} \quad W_1 \wedge W_2 \wedge \cdots \wedge W_{3g-3}$$

is a local section of the canonical bundle $\mathcal{K}_{\mathcal{M}_g}$. If $\omega_1, \omega_2, \ldots, \omega_g$ is a basis of the global differentials on the curve C (depending holomorphically on the local coordinates on $\mathcal{M}_g$) then

$$\omega_1 \wedge \omega_2 \wedge \cdots \wedge \omega_g$$

is a local section of the Hodge bundle λ. If we choose an open neighbourhood U of the point p where all w_i and ω_j are defined and their collection is still a basis of the differentials, resp. quadratic differentials, then

$$s := \frac{w_1 \wedge w_2 \wedge \cdots \wedge w_{3g-3}}{(\omega_1 \wedge \omega_2 \wedge \cdots \wedge \omega_g)^{13}}$$

is a nowhere vanishing local section over U of the trivial bundle

$$\mathcal{K}_{\mathcal{M}_g} \otimes \lambda^{-13}.$$

If we compare it with our global section μ_g we see that there is a holomorphic function F_U such that

$$\mu_g = F_U \cdot s.$$

Of course F_U depends on the choice of our basis w_i and ω_j. The Belavin–Knizhnik theorem [BK] says that the Polyakov form (the integrand of the above partition function) is

$$d\pi_g = \text{const} |F_U|^2 (-i)^g \frac{W_1 \wedge \overline{W}_1 \wedge W_2 \wedge \overline{W}_2 \wedge \cdots \wedge W_{3g-3} \wedge \overline{W}_{3g-3}}{|\det(\int \omega_i \wedge \omega_j)|^{13}}.$$

Here $\int$ means integration over the curve C. In the Beilinson–Manin article [BM] there is an explicit description of the local functions F_U if one uses certain distinguished bases.

10.2 The Grothendieck–Riemann–Roch Theorem

Before formulating the Grothendieck–Riemann–Roch theorem (GRR) we have to introduce two new objects for every nonsingular quasi-projective variety X, the Grothendieck group $K(X)$ and the Chow ring $CH(X)$. Both objects are useful in many respects.

Let us consider first $K(X)$. We take the set of isomorphy classes of vector bundles and the free abelian group F generated by this set. (See Sect. 3.2, where we used a free abelian group to define the group of divisors.) This group is too big. We have to take the factor group of this group with respect to the subgroup U generated by the elements which are described as follows. If

$$0 \rightarrow E_1 \rightarrow E \rightarrow E_2 \rightarrow 0$$

is a short exact sequence of vector bundles then we take the element

$$E - E_1 - E_2$$

in F as an element of the generating set of the subgroup U. We define the *Grothendieck group* by

$$K(X) := \frac{F}{U}.$$

We use for short: the Grothendieck group is generated by the isomorphy classes of vector bundles modulo short exact sequences.

We can substitute the word vector bundle above by coherent sheaves and again construct the Grothendieck group $K_0(X)$. But in fact in the case of nonsingular projective varieties both groups are isomorphic.[5] The reason for this is that for every coherent sheaf $\mathcal{G}$, there exists a *locally free resolution*

$$0 \rightarrow L_n \rightarrow L_{n-1} \rightarrow \cdots \rightarrow L_0 \rightarrow \mathcal{G} \rightarrow 0,$$

which says that the above sequence is exact and the L_i are locally free. Now ([...] denotes the element in the K group)

$$[\mathcal{G}] \rightarrow \sum_{i=0}^{n} (-1)^i [L_i]$$

gives the inverse map $K_0(X) \rightarrow K(X)$ to the obvious map $K(X) \rightarrow K_0(X)$.

Let $\pi : X \rightarrow Y$ be a proper algebraic map, X and Y quasi-projective and nonsingular. (*Proper* means that for A a compact set in Y, $\pi^{-1}(A)$ is a

---

[5] See Borel/Serre, [BS], p. 105.

compact set.) Typical examples of these are our families of curves of genus g with nonsingular base variety. We want to define a map from $K(X)$ to $K(Y)$ which represents the process of taking the direct image. This is done by

$$\pi_! : K(X) \to K(Y), \quad \pi_!([E]) = \sum_i (-1)^i [R^i \pi_* E].$$

The fact that this is well defined is not trivial at all. For example, a short exact sequence of bundles over X does not yield a short exact sequence of the direct image bundles. Instead there is a long exact sequence (similar to the cohomology sequence) involving all higher direct images. Also in general the higher direct images are not necessarily locally free. Here one has to use the group $K_0(X)$. Because this does not occur in our situation we will not discuss it in detail.

Having defined the Grothendieck group we come to the *Chow ring* $\mathrm{CH}(X)$. $\mathrm{CH}^i(X)$ is the free abelian group generated by irreducible closed subvarieties of X of codimension i modulo *rational equivalence*. Because we will need in the following only CH^0 and CH^1, we will not define rational equivalence here.[6] Remember that the divisor group is also generated by the irreducible subvarieties of codimension 1. Rational equivalence in codimension 1 is the same as linear equivalence of divisors. Hence

$$\mathrm{CH}^1(X) = \mathrm{Pic}\, X, \quad \mathrm{CH}^0(X) = \mathbb{Z} \cdot [X].$$

The multiplicative structure is induced by the "intersection" of subvarieties. Let

$$[A] \in \mathrm{CH}^i(X), \quad [B] \in \mathrm{CH}^j(X)$$

be classes which are represented by geometric subvarieties A and B, which cut each other transversely. Then $A \cap B$ is a subvariety of codimension $(i + j)$. The multiplication is defined by

$$[A] \cdot [B] = [A \cap B] \in \mathrm{CH}^{i+j}(X).$$

This rule induces a multiplication on the whole of $\mathrm{CH}(X)$.

If E is a vector bundle of rank k, then we can assign to E its *Chern classes*

$$c_i(E) \in \mathrm{CH}^i(X), \quad i = 0, 1, \ldots, k. \tag{10.3}$$

Because we need only c_0 and c_1 we will not give the general definition. We define

$$c_0(E) = [X] \quad \text{and} \quad c_1(E) = c_1(\wedge^k E)$$

where the c_1 on the right-hand side is the Chern class of line bundles, which is defined as the corresponding divisor class of the line bundle, see Proposition 9.3. Note that the Chern number (8.6) is the degree of this divisor.

---

[6] See Hartshorne, [Ha], App. A for this and the following results.

Let $\pi : X \to Y$ be a proper map as above, then again there is a map

$$\pi_* : \mathrm{CH}(X) \to \mathrm{CH}(Y).$$

It is induced by the following map on the irreducible closed subvarieties. If A is a subvariety of dimension k, then the dimension of $\pi(A)$ is $\leq k$. If it is strictly less than k, we set $\pi_*([A]) = 0$. If it is equal to k, we set

$$\pi_*([A]) = n_A \cdot [\overline{\pi(A)}].$$

Here $\overline{\pi(A)}$ denotes the closure of $\pi(A)$ and n_A is a positive number. If A is a point, then $n_A = 1$. For general A (which is not important for us) we take as n_A the degree of the field extension of the field of meromorphic functions on A over the field of meromorphic functions on $\pi(A)$. It is clear that

$$\pi_* : \mathrm{CH}^i(X) \to \mathrm{CH}^{i-l}(Y)$$

is a group homomorphism $(l = \dim X - \dim Y)$.

We have two important maps

$$ch, \, td : \quad \mathrm{K}(X) \quad \to \quad \mathrm{CH}(X) \otimes \mathbb{Q}$$

which are defined in terms of Chern classes. $- \otimes \mathbb{Q}$ says that we allow the coefficients to have fractional numbers. With this we lose the information in the groups $\mathrm{CH}^i(X)$ which comes from the torsion elements.

The first map is the *Chern character*

$$ch(E) := \mathrm{rank}\, E + c_1(E) + \frac{1}{2}(c_1(E)^2 - 2c_2(E)) + \text{higher order terms.} \quad (10.4)$$

"Higher order terms" means elements of $\mathrm{CH}^j(X)$ with $j > 2$.[7] We use E instead of $[E]$ and set $c_j(E) = 0$ for $j > \mathrm{rank}\, E$. The Grothendieck group $\mathrm{K}(X)$ becomes a ring if we use as multiplication the tensor product of vector bundles. With respect to the product the Chern character ch is a ring homomorphism, i.e.

$$ch(E \otimes F) = ch(E) \cdot ch(F), \qquad ch(E \oplus F) = ch(E) + ch(F).$$

The second map is the *Todd class*

$$td(E) := 1 + \frac{1}{2}c_1(E) + \frac{1}{12}(c_1(E)^2 + c_2(E)) + \text{higher order terms.} \quad (10.5)$$

It fulfils the following relation:

$$td(E \oplus F) = td(E) \cdot td(F).$$

If we take as bundle E the (holomorphic) tangent bundle of the variety Z then we just write $td(Z)$.

---

[7] See Hartshorne, [Ha], p. 432 for the general definition of all terms.

Theorem 10.4. *(Grothendieck–Riemann–Roch (GRR))* Let X and Y be non-singular quasi-projective varieties, $\pi : X \to Y$ a proper algebraic map, then the following diagram

$$K(X) \xrightarrow{ch(-)\cdot td(X)} CH(X) \otimes \mathbb{Q}$$

$$\downarrow \pi_! \qquad\qquad\qquad \downarrow \pi_*$$

$$K(Y) \xrightarrow{ch(-)\cdot td(Y)} CH(Y) \otimes \mathbb{Q}$$

commutes, i.e. for a locally free sheaf E, we have

$$ch(\pi_! E) \cdot td(Y) = \pi_*(ch(E).td(X)).$$

If $\pi : X \to B$ is a family of nonsingular curves over a nonsingular base variety and E is a locally free sheaf we get

$$ch(\pi_! E) = \pi_*(ch(E).td(\omega^*)),$$

where ω is the sheaf of the relative differentials of the family X.

Proofs can be found, e.g. in [BS], [Fu].

Now let $\pi : X \to B$ be a family of nonsingular curves over a nonsingular base variety and L a line bundle. We get

$$ch(\pi_! L) = \pi_*\left(\left(1 + c_1(L) + \frac{1}{2}c_1(L)^2 + \cdots\right)\cdot\left(1 + \frac{1}{2}c_1(\omega^*) + \frac{1}{12}c_1(\omega^*)^2 + \cdots\right)\right).$$

We have (9.1) $c_1(M^*) = -c_1(M)$, hence

$$c_1(\omega^*) = -c_1(\omega), \qquad c_1(\omega^*)^2 = c_1(\omega)^2.$$

If we multiply and order with respect to the codimension we find

$$\mathrm{rank}(\pi_! L) + c_1(\pi_! L) + \cdots = \pi_*\left(1 + \left(c_1(L) - \frac{1}{2}c_1(\omega)\right)\right.$$

$$\left. + \left(\frac{1}{2}c_1(L)^2 + \frac{1}{12}c_1(\omega)^2 - \frac{1}{2}c_1(L)c_1(\omega)\right) + \cdots\right).$$

Because π_* either kills a subvariety or leaves its dimension invariant and $\dim X = \dim B + 1$ we see that π_* maps

$$\pi_{*|} : CH^i(X) \otimes \mathbb{Q} \to CH^{i-1}(B) \otimes \mathbb{Q}, \quad i = 1, 2, \ldots.$$

Considering only the elements of $CH^1(B) \otimes \mathbb{Q}$ we get

$$c_1(\pi_* L - R^1 \pi_* L) = c_1(\pi_* L) - c_1(R^1 \pi_* L)$$

$$= \pi_* \left(\frac{1}{2} c_1(L)^2 + \frac{1}{12} c_1(\omega)^2 - \frac{1}{2} c_1(L) c_1(\omega) \right).$$

If we take $L = \omega$ then we saw (10.1) that $R^1 \pi_* \omega = \mathcal{O}_B$, and hence

$$c_1(R^1 \pi_* \omega) = c_1(\mathcal{O}_B) = 0,$$

$$c_1(\pi_* \omega) = c_1(\wedge^g \pi_* \omega) = \lambda_1 = \pi_* \left(\frac{1}{12} c_1(\omega)^2 \right) = \frac{1}{12} \pi_* (c_1(\omega)^2).$$

If we take $L = \omega^n$, $n \geq 2$ then we get $R^1 \pi_* \omega^n = 0$ (10.2) and

$$\lambda_n = c_1(\wedge^{\cdots} \pi_* (\omega^n)) = c_1(\pi_* (\omega^n))$$

$$= \pi_* \left(\frac{1}{2} c_1(\omega^n)^2 + \frac{1}{12} c_1(\omega)^2 - \frac{1}{2} c_1(\omega^n) c_1(\omega) \right)$$

$$= \pi_* \left(\left(\frac{1}{2} n^2 + \frac{1}{12} - \frac{n}{2} \right) c_1(\omega)^2 \right) = \frac{1}{12} (6n^2 - 6n + 1) \pi_* (c_1(\omega)^2).$$

(Here we used $c_1(\omega^n) = n \cdot c_1(\omega)$.)
If we compare the expressions for λ_1 and λ_n we get the relation

$$\lambda_n = (6n^2 - 6n + 1) \cdot \lambda_1.$$

Of course we have shown this only up to torsion in the Picard group of B. Some finer analysis shows that it is valid even in the Picard group.[8] Reformulated in terms of line bundles

Proposition 10.5.

$$\boxed{\lambda_n \cong \lambda_1^{6n^2 - 6n + 1}}$$

and especially for $n = 2$

$$\lambda_2 \cong \lambda_1^{13}.$$

The above isomorphism is the Mumford isomorphism already formulated in Theorem 10.1.

After having formulated the GRR theorem, I would like to show that the usual Riemann–Roch theorem is an easy consequence of it.

Let X be a nonsingular, projective variety of dimension n, E a (holomorphic, algebraic) vector bundle over X and $pt = \{*\}$ the variety consisting of one point. The constant map

$$\pi : X \to pt$$

is a proper map. The vector bundles over pt are the (finite-dimensional) complex vector spaces. Now

---

[8] Mumford, [M], Thm. 5.10, essentially the argument is that the Picard group of the moduli functor is torsion free.

$$R^i \pi_* E(pt) = \mathrm{H}^i(\pi^{-1}(pt), E_{|\pi^{-1}(pt)}) = \mathrm{H}^i(X, E)$$

and hence the higher direct image sheaves of E are the cohomology groups of E. The Chow ring of a point is calculated easily as $\mathrm{CH}(pt) = \mathbb{Z} \cdot [pt]$. If F is a vector space (hence a vector bundle on pt) then $ch(F) = \dim F \cdot [pt]$. The tangent bundle at a point is the zero bundle, hence $td(pt) = [pt]$. Theorem 10.4 yields

$$\left(\sum_{i=0}^{n}(-1)^i \dim \mathrm{H}^i(X, E)\right) \cdot [pt] = ch\left(\sum_{i=0}^{n}(-1)^i[\mathrm{H}^i(X, E)]\right) = \pi_*(ch(E).td(X)).$$
$$(10.6)$$

Now

$$\pi_{*|} : \mathrm{CH}^n(X) \to \mathrm{CH}^0(pt)$$

is the only nonvanishing part of π_*, and this map is induced by mapping a point on X to the point pt. If $[D]$ is an element of $\mathrm{CH}^n(X)$ it can be represented by a finite sum of points

$$D = \sum_{p \in X} n_p \cdot p.$$

Now

$$\deg[D] := \deg D := \sum_{p \in X} n_p$$

is well defined. We can describe π_* as

$$\pi_*(D) = (\deg D) \cdot [pt] .$$

We reformulate (10.6) as the famous

Theorem 10.6. *(Hirzebruch–Riemann–Roch)*

$$\sum_{i=0}^{n}(-1)^i \dim \mathrm{H}^i(X, E) = \deg\left((ch(E) \cdot td(X))_n\right). \qquad (10.7)$$

The right-hand side says: calculate $ch(E).td(X)$ in the Chow ring of X, take the part of the result which lies in $\mathrm{CH}^n(X)$ and calculate its degree.

We get our Riemann–Roch theorem by further specialization of Theorem 10.6. If X is a projective, nonsingular curve C, E a rank k bundle then

$$ch(E) = k + c_1(\wedge^k E).$$

If T is the tangent line bundle then

$$td(C) = 1 + \frac{1}{2}c_1(T).$$

Now $c_1(T) = -K$ where K is the canonical class. We calculate

$$\dim \mathrm{H}^0(C, E) - \dim \mathrm{H}^1(C, E) = \deg \left((k + c_1(\wedge^k E)) \left(1 - \frac{1}{2}K \right) \right)_1$$

$$= \deg \left(-\frac{k}{2}K + c_1(\wedge^k E) \right) \tag{10.8}$$

$$= k(1 - g) + \deg(c_1(\wedge^k E)) \ .$$

This is the Riemann–Roch theorem for vector bundles on projective curves. If E is a line bundle L we get simply

$$\dim \mathrm{H}^0(C, L) - \dim \mathrm{H}^1(C, L) = \deg L - g + 1,$$

which is the form of Theorem 9.6.

One final remark: In the analytic context we can substitute the Chow ring by the cohomology ring, generated by the de Rham classes of differentials on X. There is always a map from the Chow ring to the cohomology ring. But the Chow ring contains much more information concerning the variety X. For the Hirzebruch–Riemann–Roch theorem this additional information is not necessary. It can be formulated and deduced by using only the cohomology ring. In this case taking the degree in Theorem 10.6 corresponds to representing the element in $\mathrm{H}^{2n}(X, \mathbb{C})$ by a differential form and integrating it over the manifold X.

Hints for Further Reading

The direct image sheaves are covered in the books

[Ha] Hartshorne, R., *Algebraic Geometry*, Springer, 1977.

[GH] Griffiths, Ph., Harris, J., *Principles of Algebraic Geometry*, Wiley, New York, 1978.

On the Chow ring, Chern classes in the algebraic setting, Grothendieck groups you find a short collection of the relevant facts in Appendix A of Hartshorne's book. The proof of the GRR theorem is not contained there. For this you should consult the very detailed expositions in

[BS] Borel, A., Serre, J.P., Le Théorème de Riemann–Roch, *Bull. Soc. Math. France* **86** (1958) 97–136.

[Fu] Fulton, W., *Intersection Theory*, Springer, 1985.

The Mumford isomorphism is covered in (Theorem 5.10)

[M] Mumford, D., Stability of Projective Varieties, *L'Enseign. Math.* **23** (1977) 39–110.

See also

[HM] Harris, J., Mumford, D., On the Kodaira Dimension of the Moduli Space of Curves, *Invent. Math.* **67** (1982) 23–86.

On its consequences on string theory you find more information in

[BM] Beilinson, A.A., Manin, Yu.I., The Mumford Form and the Polyakov Measure in String Theory, *Commun. Math. Phys.* **107** (1986) 359–376.

[BK] Belavin, A.A., Knizhnik, V.G., Complex Geometry and Theory of Quantum Strings, *Sov. Phys. JETP* **64** (1987) 214.

[Bo] Bost, J.B., Fibrés déterminants, déterminants régularisés et mesures sur les espace de modules des courbes complexes, Sém. Bourbaki Exp. 676, *Asterisque* **152–153** (1987) 113–149.

[Co] Cornalba, M., The Belavin-Knizhnik-Bost-Jolicoeur-Gava-Iengo etc. Theorem, Summer Workshop on High Energy Physics and Cosmology, 30 June–15 August 1986, ICTP, Trieste.

[Sm] Smit, D.-J., *String Theory and Algebraic Geometry of Moduli Spaces*, Comm.Math.phys, 114 (1988), 645–685.

[Al] Alvarez-Gaumé, L., Nelson, P., Riemann Surfaces and String Theories, (in) *Supersymmetry, Supergravity and Superstrings 86*, World Scientific, Singapore, 1986.

The relation between K-groups and the cohomology ring is also used in the Atiyah-Singer index theorem. See

[Gi] Gilkey, P.B., *Invariance Theory, The Heat Equation and the Atiyah-Singer Index Theorem*, Publish or Perish, Houston, 1984.

11

Modern Algebraic Geometry

In the following two sections we will describe the modern language and the modern point of view of algebraic geometry. One of the basic ideas is to replace the classical geometric space M by the functions on the space M or, even more general, by the set of maps from this space to other spaces and vice versa. The geometry of M corresponds to some algebraic structure of the set of maps. These concepts were established by Grothendieck in the 1960s. What does one gain by this generalization? One of the advantages is that it is possible to extend the techniques of geometry to more general objects which do not have an obvious structure of a "usual space".

This point of view was very fruitful in mathematics. I do not want to give an historic account of this claim. Let me just mention the proof of Weil conjectures by Pierre Deligne, Faltings' proof of Mordell's conjecture, Faltings' proof of the Verlinde formula, Wiles' proof of Fermat's Last Theorem, etc. But it is also of importance in Theoretical Physics. There noncommutative spaces, quantum groups, etc. show up. This is one of the reasons for the increasing interest in modern algebraic geometry among theoretical physicists. Moreover from the problem of the existence of moduli spaces one is led immediately outside of the category of "usual varieties". Also singular spaces are already incorporated from the very beginning.

Starting from varieties (already introduced in Chap. 6 and Sect. 8.3, but their definition will be recalled) we will discuss the Zariski topology in detail, introduce the spectrum of a ring, show how the concept of a point can be replaced by the concept of a homomorphism. Also the non-commutative case, of relevance for the quantum plane and for quantum groups, will be touched. In this section we mainly deal with the local aspects. In the next section global aspects, i.e. affine schemes and general schemes, will be introduced.

11.1 Varieties

In classical algebraic geometry the subsets defining the geometry are the set of points where a given set of polynomials have a common zero (if we plug

M. Schlichenmaier, Modern Algebraic Geometry. In: M. Schlichenmaier, An Introduction to Riemann Surfaces, Algebraic Curves and Moduli Spaces, Theoretical and Mathematical Physics, 133–154 (2007)
DOI 10.1007/978-3-540-71175-9_12

in the coordinates of these points into the polynomials). To give an example, the polynomials X and Y are elements of the polynomial ring in two variables over the real numbers $\mathbb{R}$. They define the following polynomial functions:

$$X, Y : \mathbb{R}^2 \to \mathbb{R}, \qquad (\alpha, \beta) \mapsto X(\alpha, \beta) = \alpha, \;\; \text{resp.} \;\; Y(\alpha, \beta) = \beta .$$

These two functions are called coordinate functions. The point $(\alpha_0, \beta_0) \in \mathbb{R}^2$ can be given as the common zero set of the polynomials $X - \alpha_0$ and $Y - \beta_0$, i.e.

$$\{(\alpha_0, \beta_0)\} = \{(\alpha, \beta) \in \mathbb{R}^2 \mid X(\alpha, \beta) - \alpha_0 = 0, \; Y(\alpha, \beta) - \beta_0 = 0 \} .$$

Let me come to the general definition. For this let $\mathbb{K}$ be an arbitrary field (e.g. $\mathbb{C}$, $\mathbb{R}$, $\mathbb{Q}$, $\mathbb{F}_p$, $\overline{\mathbb{F}}_p, \ldots$) and

$$\mathbb{K}^n = \underbrace{\mathbb{K} \times \mathbb{K} \times \cdots \times \mathbb{K}}_{n \text{ times}}, \tag{11.1}$$

the n-dimensional affine space over $\mathbb{K}$, sometimes also denoted by $\mathbb{A}^n(\mathbb{K})$. Let $R_n = \mathbb{K}[X_1, X_2, \ldots, X_n]$ be the polynomial ring in n variables. A subset A of $\mathbb{K}^n$ should be a geometric object if there exist finitely many polynomials $f_1, f_2, \ldots, f_s \in R_n$ such that

$$x \in A \quad \text{if and only if} \quad f_1(x) = f_2(x) = \cdots = f_s(x) = 0.$$

Here and in the following it is understood that $x = (x_1, x_2, \ldots, x_n) \in \mathbb{K}^n$ and $f(x) \in \mathbb{K}$ denotes the element of $\mathbb{K}$ obtained by replacing the variable X_1 by the coordinate x_1, etc.

Using the notion of ideals it is possible to define these sets A in a more elegant way. An ideal of an arbitrary ring R is a subset of R which is closed under addition: $I + I \subseteq I$, and under multiplication with the whole ring: $R \cdot I \subseteq I$. A good reference to recall the necessary prerequisites from algebra is [Ku]. Now let $I = (f_1, f_2, \ldots, f_s)$ be the ideal generated by the polynomials $f_1, f_2, \ldots, f_s$ which define A, e.g.

$$I = R \cdot f_1 + R \cdot f_2 + \cdots + R \cdot f_s = \{r_1 f_1 + r_2 f_2 + \cdots + r_s f_s \mid r_i \in R, \; i = 1, \ldots, s\} .$$

Definition 11.1. *A subset A of $\mathbb{K}^n$ is called an* algebraic set *if there is an ideal I of R_n such that*

$$x \in A \iff f(x) = 0 \quad \text{for all } f \in I.$$

The set A is called the vanishing set *of the ideal I, in symbols $A = V(I)$ with*

$$V(I) := \{ x \in \mathbb{K}^n \mid f(x) = 0, \; \forall f \in I \} . \tag{11.2}$$

Remark 11.2. It is enough to test the vanishing with respect to the generators of the ideal in the definition.

Remark 11.3. There is no finiteness condition mentioned in the definition. Indeed this is not necessary, because the polynomial ring R_n is a noetherian ring. Recall a ring is a *noetherian ring* if every ideal has a finite set of generators. There are other useful equivalent definitions of a noetherian ring. Let me here recall only the fact that every strictly ascending chain of ideals (starting from one ideal) consists only of finitely many ideals. A field $\mathbb{K}$ has only the (trivial) ideals $\{0\}$ and $\mathbb{K}$, hence $\mathbb{K}$ is noetherian. Trivially all principal ideal rings (i.e. rings where every ideal can be generated by just one element) are noetherian. Besides the fields there are two important examples of principal ideal rings: $\mathbb{Z}$ the integers and $\mathbb{K}[X]$ the polynomial ring in one variable over the field $\mathbb{K}$.

Let me recall the proof for $\mathbb{Z}$. Take I an ideal of $\mathbb{Z}$. If $I = \{0\}$ we are done. Hence assume $I \neq \{0\}$, then there is an $n \in \mathbb{N}$ with $n \in I$ minimal. We now claim $I = (n)$. To see this take $m \in I$. By the division algorithm of Euclid there are $q, r \in \mathbb{Z}$ with $0 \leq r < n$ such that $m = qn + r$. Hence with m and n in I we get $r = m - qn \in I$. But n was chosen minimal, hence $r = 0$ and $m \in (n)$. Note that the proof for $\mathbb{K}[X]$ is completely analogous if we replace the division algorithm for the integers by the division algorithm for polynomials.

Now we have

Theorem 11.4. *(Hilbertscher Basissatz) Let R be a noetherian ring. Then $R[X]$ is also noetherian.*

As a nice exercise you may try to prove it by yourself (maybe guided by [Ku]).

Remark 11.5. If R is a noncommutative ring one has to deal with left, right and two-sided ideals. It is also necessary to define left, right and two-sided noetherian.

It is time to give some *examples of algebraic sets*:
(1) The whole affine space is the zero set of the zero ideal: $\mathbb{K}^n = V(0)$.
(2) The empty set is the zero set of the whole ring R_n: $\emptyset = V((1))$.
(3) Let $\alpha = (\alpha_1, \alpha_2, \ldots, \alpha_n) \in \mathbb{K}^n$ be a point given by its coordinates. Define the ideal

$$I_\alpha = (X_1 - \alpha_1, X_2 - \alpha_2, \ldots, X_n - \alpha_n),$$

then $\{\alpha\} = V(I_\alpha)$.
(4) Now take two points α, β and their associated ideals I_α, I_β as defined in (3).
Then $I_\alpha \cap I_\beta$ is again an ideal and we get $\{\alpha, \beta\} = V(I_\alpha \cap I_\beta)$.
This is a *general fact*.

Proposition 11.6. *Let $A = V(I)$ and $B = V(J)$ be two algebraic sets then the union $A \cup B$ is again an algebraic set, i.e. $A \cup B = V(I \cap J)$, i.e.*

$$V(I) \cup V(J) \quad = \quad V(I \cap J). \tag{11.3}$$

Proof. Obviously we get for two ideals K and L with $K \subseteq L$ for their vanishing sets $V(K) \supseteq V(L)$. As $I \cap J \subseteq I$ and $I \cap J \subseteq J$ we obtain $V(I \cap J) \supseteq V(I) \cup V(J)$. To show the other inclusion, take an $x \notin V(I) \cup V(J)$ then there are $f \in I$ and $g \in J$ with $f(x) \neq 0$ and $g(x) \neq 0$. Now $f \cdot g \in I \cap J$ but $(f \cdot g)(x) = f(x) \cdot g(x) \neq 0$. Hence $x \notin V(I \cap J)$. $\square$

(**5**) A hypersurface H is the vanishing set of the ideal generated by a single polynomial f: $H = V((f))$. In $\mathbb{C}^2$ an example is given by $I = (Y^2 - 4X^3 + g_2 X + g_3)$ where $g_2, g_3 \in \mathbb{C}$. The set $V(I)$ defines a cubic curve in the plane. For general g_2, g_3 this curve is isomorphic to a (complex) one-dimensional torus with the point 0 removed, see Sect. 6.2.

(**6**) Linear affine subspaces are algebraic sets. A linear affine subspace of $\mathbb{K}^n$ is the set of solutions of a system of linear equations $A \cdot x = \mathbf{b}$ with

$$A = \begin{pmatrix} \mathbf{a}_{1,*} \\ \cdots \\ \mathbf{a}_{r,*} \end{pmatrix}, \quad b = \begin{pmatrix} b_1 \\ \cdots \\ b_r \end{pmatrix}, \quad \mathbf{a}_{i,*} \in \mathbb{K}^n, \quad b_i \in \mathbb{K}, i = 1, \ldots, r.$$

The solutions (by definition) are given as the elements of the vanishing set of the ideal

$$I = (\mathbf{a}_{1,*} \cdot X - b_1, \ \mathbf{a}_{2,*} \cdot X - b_2, \ \cdots, \ \mathbf{a}_{r,*} \cdot X - b_r),$$

with $X = {}^t(X_1, X_2, \ldots, X_n)$ a vector of variables.

(**7**) A special case are the straight lines in the plane. For this let

$$l_i = a_{i,1} X + a_{i,2} Y - b_i, \quad i = 1, 2$$

be two linear forms. Then $L_i = V((l_i))$, $i = 1, 2$ are lines. For the union of the two lines we obtain by (11.3)

$$L_1 \cup L_2 = V((l_1) \cap (l_2)) = V((l_1 \cdot l_2)).$$

Note that I do not claim $(l_1) \cap (l_2) = (l_1 \cdot l_2)$. The reader is encouraged to search for conditions when this will hold. For the intersection of the two lines we get $L_1 \cap L_2 = V((l_1, l_2))$ which can be written as $V((l_1) + (l_2))$. Of course this set consists just of one point if the linear forms l_1 and l_2 are linearly independent. Again, there is the *general fact*

$$V(I) \cap V(J) = V(I + J), \tag{11.4}$$

where

$$I + J := \{ f + g \mid f \in I, g \in J \}$$

is the sum of two ideals.

You see there is an ample supply of examples for algebraic sets.

Next we introduce for $\mathbb{K}^n$ a topology, the *Zariski topology*. For this we call a subset U open if it is a complement of an algebraic set, i.e. $U = \mathbb{K}^n \setminus V(I)$ where I is an ideal of R_n. In other words, the closed sets are the algebraic sets. It is easy to verify the axioms for a topology:

(1) $\mathbb{K}^n$ and $\emptyset$ are open.
(2) Finite intersections are open:

$$U_1 \cap U_2 = (\mathbb{K}^n \backslash V(I_1)) \cap (\mathbb{K}^n \backslash V(I_2)) = \mathbb{K}^n \backslash (V(I_1) \cup V(I_2)) = \mathbb{K}^n \backslash V(I_1 \cap I_2).$$

(3) Arbitrary unions are open:

$$\bigcup_{i \in S} (\mathbb{K}^n \backslash V(I_i)) = \mathbb{K}^n \backslash \bigcap_{i \in S} V(I_i) = \mathbb{K}^n \backslash V\left(\sum_{i \in S} I_i\right).$$

Here S is allowed to be an infinite index set. The ideal $\sum_{i \in S} I_i$ consists of elements in R_n which are finite sums of elements belonging to different I_i. Equation (11.4) easily extends to this setting.

Let us study the affine line $\mathbb{K}$. Here $R_1 = \mathbb{K}[X]$. All ideals in $\mathbb{K}[X]$ are principal ideals, i.e. generated by just one polynomial. The vanishing set of an ideal consists just of the finitely many zeros of this polynomial (if it is not identically zero). Conversely, for every set of finitely many points there is a polynomial vanishing exactly at these points. Hence, besides the empty set and the whole line the algebraic sets are the sets of finitely many points. At this level there is already a new concept showing up. The polynomial assigned to a certain point set is not unique. For example it is possible to increase the vanishing order of the polynomial at a certain zero without changing the vanishing set. It would be better to talk about point sets with multiplicities to get a closer correspondence to the polynomials. Additionally, if $\mathbb{K}$ is not algebraically closed then there are nontrivial polynomials without any zeros at all. We will take up these ideas later. The other important observation is that the open sets in $\mathbb{K}$ are either empty or dense. The latter says that for nonempty U the closure $\overline{U}$ of U, i.e. the smallest closed set which contains U, is the whole space $\mathbb{K}$. Assuming the whole space to be irreducible this is true in a more general context.

Definition 11.7.
(a) Let V be a closed set. V is called irreducible if for every decomposition $V = V_1 \cup V_2$ with V_1, V_2 closed we have $V_1 = V$ or $V_2 = V$.
(b) An algebraic set which is irreducible is called a variety.

Now let U be an open subset of an irreducible V. The two set $V \backslash U$ and $\overline{U}$ are closed and $V = (V \backslash U) \cup \overline{U}$. Hence V has to be one of these sets. Hence either $U = \emptyset$ or $V = \overline{U}$. As promised, this shows that every open subset of an irreducible space is either empty or dense. Note that this has nothing to do with our special situation. It follows from general topological arguments. Further down we will see that the spaces $\mathbb{K}^n$ are irreducible.

Remark 11.8. We had already introduced irreducibility in Sect. 6.1. There we called every algebraic set a variety. In view of the algebraic consequences we are considering here it is more convenient to reserve the word varieties for irreducible ones in this and the next section.

Up to now we were able to describe our geometric objects with the help of the ring of polynomials. This ring plays another important role in the whole theory. We need it to study polynomial (algebraic) functions on $\mathbb{K}^n$. If $f \in R_n$ is a polynomial then $x \mapsto f(x)$ defines a map from $\mathbb{K}^n$ to $\mathbb{K}$. This can be extended to functions on algebraic sets $A = V(I)$. We associate to A the quotient ring

$$R(A) := \mathbb{K}[X_1, X_2, \ldots, X_n]/I .$$

This ring is called the coordinate ring of A. The elements of $R(A)$ can be considered as functions on A. Take $x \in A$, and $\bar{f} := f \bmod I \in R(A)$ then $\bar{f}(x) := f(x)$ is a well-defined element of $\mathbb{K}$. Assume $\bar{f} = \bar{g}$ then there is an $h \in I$ with $f = g + h$, hence $f(x) = g(x) + h(x) = g(x) + 0$. You might have noticed that it is not really correct to call this ring the coordinate ring of A. It is not clear, in fact it is not even true, that the ideal I is fixed by the set A. But $R(A)$ depends on I. A first way to avoid these complications is to assign to every A a unique defining ideal,

$$I(A) := \{f \in R_n \mid f(x) = 0, \forall x \in A\} . \tag{11.5}$$

It is the largest ideal which defines A. For arbitrary ideals we always obtain $I(V(I)) \supseteq I$.

There is a second possibility which even takes advantage out of the nonuniqueness. We could have added the additional data of the defining ideal I in the notation. Just simply assume that when we use A it comes with a certain I. Compare this with the situation above where we determined the closed sets of $\mathbb{K}$. Again at the first glance this annoying fact of nonuniqueness of I will allow us to introduce multiplicities in the following which in turn will be rather useful as we will see.

Here another warning is in order. The elements of $R(A)$ define usual functions on the set A. But different elements can define the same function. In particular, $R(A)$ can have zero divisors and nilpotent elements (which always give the zero function).

The ring $R(A)$ contains all the geometry of A. As an example, take A to be a curve in the plane and P a point in the plane. Then $A = V((f))$ with f a polynomial in X and Y and $P = V((X - \alpha, Y - \beta))$. Now $P \subset A$ (which says that the point P lies on A) if and only if $(X - \alpha, Y - \beta) \supset (f)$. Moreover in this case we obtain the following diagram of ring homomorphisms:

$$
\begin{array}{ccc}
R_2 & \xrightarrow{\;/(f)\;} & R(A) \\[4pt]
\subseteq\uparrow & & \subseteq\uparrow \\[4pt]
(X - \alpha, Y - \beta) & \xrightarrow{\;/(f)\;} & (X - \alpha, Y - \beta)/(f) \\[4pt]
\subseteq\uparrow & & \subseteq\uparrow \\[4pt]
(f) & \xrightarrow{\;/(f)\;} & \{0\} .
\end{array}
$$

The quotient $(X - \alpha, Y - \beta)/(f)$ is an ideal of $R(A)$ and corresponds to the point P lying on A.

Indeed, this is the general situation which we will study in the following sections: the algebraic sets on A correspond to the ideals of $R(A)$ which in turn correspond to the ideals lying between the defining ideal of A and the whole ring R_n.

Let me close this subsection by studying the geometry of a single point $P = (\alpha, \beta) \in \mathbb{K}^2$. A defining ideal is $I = (X - \alpha, Y - \beta)$. If we require "multiplicity one" this is the defining ideal. Hence the coordinate ring $R(P)$ of a point is $\mathbb{K}[X, Y]/I \cong \mathbb{K}$. The isomorphism is induced by the homomorphism $\mathbb{K}[X, Y] \to \mathbb{K}$ given by $X \to \alpha$, $Y \to \beta$. Indeed, every element r of $\mathbb{K}[X, Y]$ can be given as

$$r = r_0 + (X - \alpha) \cdot g + (Y - \beta) \cdot f, \quad r_0 \in \mathbb{K}, \ f, g \in \mathbb{K}[X, Y] . \tag{11.6}$$

Under the homomorphism r maps to r_0. Hence r is in the kernel of the map if and only if r_0 equals 0 which in turn is the case if and only if r is in the ideal I. The description (11.6) also shows that I is a maximal ideal. We call an ideal I a maximal ideal if there are no ideals between I and the whole ring R (and $I \neq R$). Any ideal strictly larger than the above I would contain an r with $r_0 \neq 0$. Now this ideal would contain $r, (X - \alpha), (Y - \beta)$, hence also r_0. Hence also $(r_0)^{-1} \cdot r_0 = 1$. But an ideal containing 1 is always the whole ring.

On the geometric side the points are the minimal sets. On the level of the ideals in R_n this corresponds to the fact that an ideal defining a point (with multiplicity one) is a maximal ideal. If the field $\mathbb{K}$ is algebraically closed then every maximal ideal corresponds indeed to a point.

11.2 The Spectrum of a Ring

In the last section we saw that geometric objects of a variety are in correspondence to algebraic objects of the coordinate ring of the variety. We will develop this more systematically in this section. We have the following correspondences (see (11.2) and (11.5))

$$\text{ideals of } R_n \xrightarrow{\ V\ } \text{algebraic sets}$$

$$\text{ideals of } R_n \xleftarrow{\ I\ } \text{algebraic sets} .$$

Recall the definitions: $(R_n = \mathbb{K}[X_1, X_2, \ldots, X_n])$

$$\begin{aligned} V(I) &:= \{ x \in \mathbb{K}^n \mid f(x) = 0, \ \forall f \in I \}, \\ I(A) &:= \{ f \in R_n \mid f(x) = 0, \ \forall x \in A \} . \end{aligned} \tag{11.7}$$

In general, $I(V(J))$ will be bigger than the ideal J. Let me give an example. Consider in $\mathbb{C}[X]$ the ideals $I_1 = (X)$ and $I_2 = (X^2)$. Then

$V(I_1) = V(I_2) = \{0\}$. Hence both ideals define the same point as vanishing set. Moreover $I(V(I_2)) = I_1$ because I_1 is a maximal ideal. If we determine the coordinate ring of the two situations we obtain for I_1 the ring $\mathbb{C}[X]/(X) \cong \mathbb{C}$. This is the expected situation because the functions on a point are just the constants. For I_2 we obtain $\mathbb{C}[X]/(X^2) \cong \mathbb{C} \oplus \mathbb{C} \cdot \epsilon$ the algebra generated by 1 and ϵ with the relation $\epsilon^2 = 0$ (X maps to ϵ). Hence there is no 1–1 correspondence between ideals and algebraic sets. If one wants such a correspondence one has to throw away the "wrong" ideals. This is in fact possible (by considering the so-called radical ideals, see the definition below). Indeed, it has some advantages to allow all ideals and to obtain in this way more general objects than the classical objects.

To give an example, take the affine real line and let $I_t = (X^2 - t^2)$ for $t \in \mathbb{R}$ be a family of ideals. The role of t is the role of a parameter one is allowed to vary. Obviously,

$$I_t = ((X - t)(X + t)) = (X - t) \cdot (X + t).$$

For $t \neq 0$ we obtain $V(I_t) = \{t, -t\}$ and for $t = 0$ we obtain $V(I_0) = \{0\}$. We see that for general values of t we get two points and for the value $t = 0$ one point. If we approach with t the value 0 the two different points $\pm t$ come closer and closer together. Now our intuition says that the limit point $t = 0$ should better be counted twice. We can make this intuition mathematically precise on the level of the coordinate rings. Here we have

$$R_t = \mathbb{R}[X]/I_t \cong \mathbb{R} \oplus \mathbb{R} \cdot \epsilon, \quad \epsilon^2 = t^2 . \tag{11.8}$$

The coordinate ring is a two-dimensional vector space over $\mathbb{R}$ which reflects the fact that we deal with two points. Everything here is also true for the exceptional value $t = 0$. Especially R_0 is again two-dimensional. This says we count the point $\{0\}$ twice. The drawback is that the interpretation of the elements of R_t as classical functions will not be possible in all cases. In our example for $t = 0$ the element $\bar{X}$ will be nonzero but $\bar{X}(0) = 0$.

For the following definitions let R be an arbitrary commutative ring with unit.

Definition 11.9. *(a) An ideal P of R is called a prime ideal if $P \neq R$ and $a \cdot b \in P$ implies $a \in P$ or $b \in P$.*
(b) An ideal M of R is called a maximal ideal if $M \neq R$ and for every ideal M' with $M' \supseteq M$ it follows that $M' = M$ or $M' = R$.
(c) Let I be an ideal. The radical of I is defined as

$$\mathrm{Rad}(I) := \{\, f \in R \mid \exists n \in \mathbb{N} : f^n \in I \,\} .$$

(d) The nil radical of the ring R is defined as $\mathrm{nil}(R) := \mathrm{Rad}(\{0\})$.
(e) A ring is called reduced if $\mathrm{nil}(R) = \{0\}$.
(f) An ideal I is called a radical ideal if $\mathrm{Rad}(I) = I$.

Starting from these definitions there are a lot of easy exercises for the reader:
(1) Let P be a prime ideal. Show R/P is a ring without zero divisor (such rings are called integral domains).
(2) Let M be a maximal ideal. Show R/M is a field.
(3) Every maximal ideal is a prime ideal.
(4) $\mathrm{Rad}(I)$ is an ideal.
(5) $\mathrm{Rad}(I)$ equals the intersection of all prime ideals containing I.
(6) $\mathrm{nil}(R/I) = \mathrm{Rad}(I)/I$ and conclude that every prime ideal is a radical ideal.
(7) $\mathrm{Rad}\, I$ is a radical ideal.

Let me return to the rings R_t defined by (11.8). The ideals I_t are not prime because neither $X + t$ nor $X - t$ are in I_t but $(X + t)(X - t) \in I_t$. In particular, R_t is not an integral domain: $(\epsilon + t)(\epsilon - t) = 0$. Let us calculate $\mathrm{nil}(R_t)$. For this we take an element $0 \neq z = a + b\epsilon$ and calculate

$$0 = (a + b\epsilon)^n = a^n + \binom{n}{1}a^{n-1}b^1\epsilon + \binom{n}{2}a^{n-2}b^2\epsilon^2 + \cdots .$$

Replacing ϵ^2 by the positive real number t^2 we obtain

$$0 = (a + b\epsilon)^n = \left(a^n + \binom{n}{2}a^{n-2}b^2t^2 + \binom{n}{4}a^{n-4}b^4t^4 + \cdots \right) +$$
$$+\epsilon\left(\binom{n}{1}a^{n-1}b^1 + \binom{n}{3}a^{n-3}b^3t^2 + \cdots \right) . \tag{11.9}$$

From this we conclude that all terms in the first and in the second sum have to vanish (all terms have the same sign). This implies $a = 0$. Regarding the last element in both sums we see that for $t \neq 0$ we get $b = 0$. Hence $\mathrm{nil}(R_t) = \{0\}$, for $t \neq 0$ and the ring R_t is reduced. For $t = 0$ the value of b is arbitrary. Hence $\mathrm{nil}(R_0) = (\epsilon)$, which says that R_0 is not a reduced ring. This is the typical situation: a nonreduced coordinate ring $R(V)$ corresponds to a variety V which should be considered with higher multiplicity.

For the polynomial ring we have the following very important result.

Theorem 11.10. *(Hilbertscher Nullstellensatz) Let I be an ideal in $R_n = \mathbb{K}[X_1, X_2, \ldots, X_n]$. If $\mathbb{K}$ is algebraically closed then $I(V(I)) = \mathrm{Rad}(I)$.*

The proof of this theorem is not easy. The main tool is the following version of the Nullstellensatz which more resembles his name:

Theorem 11.11. *(Hilbertscher Nullstellensatz) Let I be an ideal in $R_n = \mathbb{K}[X_1, X_2, \ldots, X_n]$ with $I \neq R_n$. If $\mathbb{K}$ is algebraically closed then $V(I) \neq \emptyset$. In other words given a set of polynomials such that the constant polynomial 1 cannot be represented as a R_n-linear sum in these polynomials then there is a simultaneous zero of these polynomials.*

For the proof let me refer to [Ku].

In the case that the field $\mathbb{K}$ is algebraically closed, the Nullstellensatz gives us a 1:1 correspondence between algebraic sets in $\mathbb{K}^n$ and the radical ideals of $R_n = \mathbb{K}[X_1, X_2, \ldots, X_n]$.

If we consider only prime ideals we get

Proposition 11.12. *Let P be a radical ideal. Then P is a prime ideal if and only if $V(P)$ is a variety, i.e. an irreducible algebraic set.*

Before we come to the proof of the proposition let me state the following simple observation. For arbitrary subsets S and T of $\mathbb{K}^n$ the ideals $I(S)$ and $I(T)$ can be defined completely in the same way as in (11.5), i.e.

$$I(S) := \{ f \in R_n \mid f(x) = 0, \forall x \in S \} . \tag{11.10}$$

It is easy to show that

$$I(S \cup T) = I(S) \cap I(T), \quad \text{and} \quad V(I(S)) = \overline{S} . \tag{11.11}$$

Here $\overline{S}$ denotes the topological closure of S, which is the smallest (Zariski-) closed subset of $\mathbb{K}^n$ containing S.

Proof. (Proposition 11.12) Let P be a prime ideal and set $Y = V(P)$ then $I(V(P)) = \mathrm{Rad}(P) = P$ by the Nullstellensatz. Assuming $Y = Y_1 \cup Y_2$ a closed decomposition of Y then $I(Y) = I(Y_1 \cup Y_2) = I(Y_1) \cap I(Y_2) = P$. Clearly $P \subseteq I(Y_1), P \subseteq I(Y_2)$. Assume $P \neq I(Y_1)$, hence there exists $a \notin P$, but $a \in I(Y_1)$. For every $b \in I(Y_2)$ we have $a \cdot b \in P$. As P is prime we get $b \in P$. This shows $I(Y_2) = P$. Altogether either $P = I(Y_1)$ or $P = I(Y_2)$. Assume the first then $Y_1 = V(I(Y_1)) = V(P) = Y$ (using that Y_1 is closed), i.e. Y is irreducible.

Conversely: let $Y = V(P)$ be irreducible with P a radical ideal. By the Nullstellensatz $P = \mathrm{Rad}(P) = I(Y)$. Let $f \cdot g \in P$ then $f \cdot g$ vanishes on Y. We can decompose

$$Y = Y \cap V(f \cdot g) = Y \cap (V(f) \cup V(g)) = (Y \cap V(f)) \cup (Y \cap V(g))$$

into closed subsets of Y. By the irreducibility it has to coincide with one of them. Assume with the first. But this implies that $V(f) \supseteq Y$ and hence f is identically zero on Y. We get $f \in I(Y) = P$. This shows that P is a prime ideal. $\square$

The fact that we restricted the situation to radical ideals corresponds to the fact that varieties as sets always have multiplicity 1, hence they are always "reduced". To incorporate all ideals and hence "nonreduced structures" we have to use the language of schemes (see Chap. 12).

Let us look at the maximal ideals of $R_n = \mathbb{K}[X_1, X_2, \ldots, X_n]$. (Still $\mathbb{K}$ is assumed to be algebraically closed.) The same argument as in the two-dimensional case shows that the ideals

$$M_\alpha = (X_1 - \alpha_1,\ X_2 - \alpha_2,\ \ldots,\ X_n - \alpha_n)$$

are maximal and that $R_n/M_\alpha \cong \mathbb{K}$. This is even true if the field $\mathbb{K}$ is not algebraically closed. Now let M' be a maximal ideal. By the Nullstellensatz (here the fact that $\mathbb{K}$ is algebraically closed is important) there is a common zero α for all elements $f \in M'$. Take $f \notin M'$ then $R_n = (f, M')$. Now $f(\alpha) = 0$ would imply that α is a zero of all polynomials in R_n which is impossible. Hence every polynomial f which vanishes at α lies in M'. All elements in M_α have α as a zero. This implies $M_\alpha \subseteq M' \subsetneq R_n$. By the maximality of M_α we conclude $M_\alpha = M'$.

Everything can be generalized to an arbitrary variety A over an algebraically closed field. The points of A correspond to the maximal ideals of R_n lying above the defining prime ideal P of A. They correspond exactly to the maximal ideals in $R(A)$. All of them can be given as M_α/P. This can be extended to the varieties of $\mathbb{K}^n$ lying on A. They correspond to the prime ideals of R_n lying between the prime ideal P and the whole ring. They in turn can be identified with the prime ideals of $R(A)$.

Coming back to arbitrary rings it is now quite useful to talk about dimensions.

Definition 11.13. *Let R be a ring. The (Krull-) dimension* $\dim R$ *of a ring R is defined as the maximal length r of all strict chains of prime ideals P_i in R*

$$P_0 \subsetneq P_1 \subsetneq P_2 \ldots \subsetneq P_r \subsetneq R\ .$$

Example 11.14. For a field $\mathbb{K}$ the only (prime) ideal is $\{0\} \subset \mathbb{K}$. Hence $\dim \mathbb{K} = 0$.

Example 11.15. The dimension of $R_n = \mathbb{K}[X_1, X_2, \ldots, X_n] = R(\mathbb{K}^n)$ is n. One should expect this result from any reasonable definition of dimension. Indeed we have the chain of prime ideals

$$(0) \subsetneq (X_1) \subsetneq (X_1, X_2) \subsetneq \cdots \subsetneq (X_1, X_2, \ldots, X_n) \subsetneq R_n\ .$$

Hence $\dim R_n \geq n$. With some more commutative algebra it is possible to show the equality, see [Ku, p. 54].

Example 11.16. As a special case one obtains $\dim \mathbb{K}[X] = 1$. Here the reason is a quite general result. Recall that $\mathbb{K}[X]$ is a principal ideal ring without zero divisors. Hence every ideal I can be generated by one element f. Assume I to be a prime ideal, $I \neq \{0\}$ and let $M = (g)$ be a maximal ideal lying above I. We show that I is already maximal. Because $(f) \subseteq (g)$ we get $f = r \cdot g$. But I is prime. This implies either r or g lies in I. If $g \in I$ we are done. If $r \in I$ then $r = s \cdot f$ and $f = f \cdot s \cdot g$. In a ring without zero divisor one is allowed to cancel common factors. We obtain $1 = s \cdot g$. Hence $1 \in M$ which contradicts the fact that M is not allowed to be the whole ring. From this it follows that

$\dim \mathbb{K}[X] = 1$. Note that we did not make any reference to the special nature of the polynomial ring here.

What are the conditions on f assuring that the ideal (f) is prime. The necessary and sufficient condition is that f is irreducible but not a unit. This says if there is decomposition $f = g \cdot h$ then either g or h has to be a unit (i.e. to be invertible) which in our situation says that g or h must be a constant. This can be seen in the following way. From the decomposition it follows (using (f) is prime) that either g or h has to be in (f) hence is a multiple of f. By considering the degree we see that the complementary factor has degree zero and hence is a constant.

Conversely, let f be irreducible but not a unit. Assume $g \cdot h \in (f)$, then $g \cdot h = f \cdot r$. In the polynomial ring we have unique factorization (up to units) into irreducible elements. Hence the factor f is contained either in g or h. This shows the claim.

Example 11.17. The ring of integers $\mathbb{Z}$ is also a principal ideal ring without zero divisor. Again we obtain $\dim \mathbb{Z} = 1$. In fact the integers behave very much (at least from the point of view of algebraic geometry) like the affine line over a field. What are the "points" of $\mathbb{Z}$? As already said the points should correspond to the maximal ideals. Every prime ideal in $\mathbb{Z}$ is maximal. An ideal (n) is prime exactly if n is a prime number. Hence the "points" of $\mathbb{Z}$ are the prime numbers.

Now we want to introduce the Zariski topology on the set of all prime ideals of a ring. First we introduce the sets

$$\text{Spec}(R) := \{ P \mid P \text{ is a prime ideal of } R \},$$
$$\text{Max}(R) := \{ P \mid P \text{ is a maximal ideal of } R \} .$$

The set $\text{Spec}(R)$ contains in some sense all irreducible "subvarieties" of the "geometric model" of R. Let S be an arbitrary subset of R. We define the associated subset of $\text{Spec}(R)$ as the set consisting of the prime ideals which contain S:

$$V(S) := \{ P \in \text{Spec}(R) \mid P \supseteq S \} . \tag{11.12}$$

The subsets of $\text{Spec}(R)$ obtained in this way are called the closed subsets. It is obvious that $S \subseteq T$ implies $V(S) \supseteq V(T)$. Clearly $V(S)$ depends only on the ideal generated by S: $V((S)) = V(S)$.

This defines a topology on $\text{Spec}(R)$ the *Zariski topology*.

(1) The whole space and the empty set are closed, as we have

$$V(0) = \text{Spec}(R) \quad \text{and} \quad V(1) = \emptyset.$$

(2) Arbitrary intersections of closed sets are again closed:

$$\bigcap_{i \in J} V(S_i) = V\left(\bigcup_{i \in J} S \right) . \tag{11.13}$$

(3) Finite unions of closed set are again closed:

$$V(S_1) \cup V(S_2) = V((S_1) \cap (S_2)) . \tag{11.14}$$

Let me just show (11.14) here. Because $(S_1), (S_2) \supseteq (S_1) \cap (S_2)$ we get $V(S_1) \cup V(S_2) \subseteq V((S_1) \cap (S_2))$. Take $P \in V((S_1) \cap (S_2))$. This says $P \supseteq (S_1) \cap (S_2)$. If $P \supseteq (S_1)$ we get $P \in V(S_1)$ and we are done. Hence assume $P \not\supseteq (S_1)$. Then there is a $y \in (S_1)$ such that $y \notin P$. But now $y \cdot (S_2)$ is a subset of both (S_1) and (S_2) because they are ideals. Hence $y \cdot (S_2) \subseteq P$. By the prime ideal condition $(S_2) \subseteq P$ which we had to show.

Remark 11.18. If we take any prime ideal P then the (topological) closure of P in $\mathrm{Spec}(R)$ is given as

$$V(P) = \{ Q \in \mathrm{Spec}(R) \mid Q \supseteq P \} .$$

Hence the closure of P consists of P and all "subvarieties" of P together. In particular the closure of a curve consists of the curve as geometric object and all points lying on the curve.

Remark 11.19. The closed points in $\mathrm{Spec}(R)$ are those prime ideals which are maximal ideals.

At the end of this section let me return to the affine line over a field $\mathbb{K}$, resp. its algebraic model, the polynomial ring in one variable $\mathbb{K}[X]$. We saw already that we have the nonclosed point corresponding to the prime ideal $\{0\}$ (also called the generic point) and the closed points corresponding to the prime ideals (f) (which are automatically maximal) where f is an irreducible polynomial of degree ≥ 1. If $\mathbb{K}$ is an algebraically closed field the only irreducible polynomials are the linear polynomials $X - \alpha$. Hence the closed points of $\mathrm{Spec}(\mathbb{K}[X])$ indeed correspond to the geometric points $\alpha \in \mathbb{K}$. The nonclosed point corresponds to the whole affine line.

Now we want to drop the condition that $\mathbb{K}$ is algebraically closed. As example let us consider $\mathbb{R}[X]$. We have two different types of irreducible polynomials. Of type (i) are the linear polynomials $X - \alpha$ (with a real zero α) and of type (ii) are the quadratic polynomials $X^2 + 2aX + b$ with pairs of conjugate complex zeros. The maximal ideals generated by the polynomials of type (i) correspond again to the geometric points of $\mathbb{R}$. There is no such relation for type (ii). In this case we have $V(X^2 + 2aX + b) = \emptyset$. Hence there is no subvariety at all associated to this ideal. But if we calculate the coordinate ring $R(A)$ of this (not existing) subvariety A we obtain

$$R(A) = \mathbb{R}[X]/(X^2 + 2aX + b) \cong \mathbb{R} \oplus \mathbb{R}\bar{X}$$

with the relation $\bar{X}^2 = -2a\bar{X} - b$. In particular, $R(A)$ is a two-dimensional vector space. It is easy to show that $R(A)$ is isomorphic to $\mathbb{C}$, if we map for example $\bar{X}$ to $-a + \sqrt{b-a^2}$ i. Instead of describing the "point" A as nonexisting we should better describe it as a point of the real affine line which is $\mathbb{C}$-valued. (Recall that for the points of type (i) $R(A) \cong \mathbb{R}$.) This corresponds to the fact that the polynomial splits over the complex numbers $\mathbb{C}$ into two factors

$$(X + (a + \sqrt{a^2 - b}))(X + (a - \sqrt{a^2 - b})).$$

In this sense, the ideals of type (ii) correspond to conjugate pairs of complex numbers. Note that there is no way to distinguish between the two conjugate numbers from our point of view.

In the general situation for $\mathbb{K}$ one has to consider $\mathbb{L}$-valued points, where $\mathbb{L}$ is allowed to be any finite-dimensional field extension of $\mathbb{K}$.

11.3 Homomorphisms

Let V and W be algebraic sets (not necessarily irreducible), and

$$R(V) = \mathbb{K}[X_1, X_2, \ldots, X_n]/I, \qquad R(W) = \mathbb{K}[Y_1, Y_2, \ldots, Y_m]/J$$

their coordinate rings. If $\Phi : V \to W$ is an arbitrary map and $f : W \to \mathbb{K}$ is a function then the pullback $\Phi^*(f) := f \circ \Phi$ is a function $V \to \mathbb{K}$. If we interpret the elements of $R(W)$ as functions we want to call Φ an *algebraic map* if $\Phi^*(f) \in R(V)$ for every $f \in R(W)$. Roughly speaking this is equivalent to the fact that Φ "comes" from an algebra homomorphism $R(W) \to R(V)$. In this sense the coordinate rings are the dual objects to the algebraic varieties.

To make this precise, especially also to take care of the multiplicities, we should start from the other direction. Let $\Psi : R(W) \to R(V)$ be an algebra homomorphism. This homomorphism defines a homomorphism $\widetilde{\Psi}$ (where ν is the natural quotient map)

$$\widetilde{\Psi} = \Psi \circ \nu : \mathbb{K}[Y_1, Y_2, \ldots, Y_m] \to R(V) \quad \text{with} \quad \widetilde{\Psi}(J) = 0 \mod I \ .$$

Such a homomorphism is given if we know the elements $\widetilde{\Psi}(Y_j)$. Conversely, if we fix elements $r_1, r_2, \ldots, r_m \in R(V)$ then $\widetilde{\Psi}(Y_j) := r_j$ for $j = 1, \ldots, m$ defines an algebra homomorphism $\widetilde{\Psi} : \mathbb{K}[Y_1, Y_2, \ldots, Y_m] \to R(V)$. If $f(r_1, r_2, \ldots, r_m) = 0 \mod I$ for all $f \in J$ then $\widetilde{\Psi}$ factorizes through $R(W)$. Such a map indeed defines a map Ψ^* on the set of geometric points,

$$\Psi^* : V \to W, \qquad \Psi^*(\alpha_1, \alpha_2, \ldots, \alpha_n) := (\beta_1, \beta_2, \ldots, \beta_m)$$

where the β_j are defined as

$$\beta_j = Y_j(\Psi^*(\alpha_1, \alpha_2, \ldots, \alpha_n)) := \widetilde{\Psi}(Y_i)(\alpha_1, \alpha_2, \ldots, \alpha_n).$$

We have to check whether $\Psi^*(\alpha) = \beta \in \mathbb{K}^m$ lies in the algebraic set W for $\alpha \in V$. For this we have to show that for all $f \in J$ we get $f(\Psi^*(\alpha)) = 0$ for $\alpha \in V$. But

$$f(\Psi^*(\alpha)) = f\big(Y_1(\Psi^*(\alpha)), \dots, Y_m(\Psi^*(\alpha))\big) =$$
$$f\big(\widetilde{\Psi}(Y_1)(\alpha), \dots, \widetilde{\Psi}(Y_m)(\alpha)\big) = \widetilde{\Psi}(f)(\alpha) \ .$$

Now $\widetilde{\Psi}(f) = 0$, hence the claim.

Example 11.20. A function $V \to \mathbb{K}$ is given on the dual objects as a $\mathbb{K}$-algebra homomorphism

$$\Phi : \mathbb{K}[T] \to R(V) = \mathbb{K}[X_1, X_2, \dots, X_n]/I.$$

Such a Φ is uniquely given by choosing an arbitrary element $a \in R(V)$ and defining $\Phi(T) := a$. Here again you see the (now complete) interpretation of the elements of $R(V)$ as functions on V.

Example 11.21. The geometric process of choosing a (closed) point α on V can alternatively be described as giving a map from the algebraic variety consisting of just one point to the variety. Changing to the dual objects such a map is given as a map Φ_α from $R(V)$ to the field $\mathbb{K}$ which is the coordinate ring of a point. In this sense points correspond to homomorphisms of the coordinate ring to the base field $\mathbb{K}$. Such a homomorphism has of course a kernel $\ker \Phi_\alpha$ which is a maximal ideal. Again it is the ideal defining the closed point α.

We will study this relation later. But first we take a different look on the situation.

For this let R be a $\mathbb{K}$-algebra where $\mathbb{K}$ is a field. The typical examples are the quotients of the polynomial ring $\mathbb{K}[X_1, X_2, \dots, X_n]$. Let M be a module over R, i.e. a linear structure over R. In particular, M is a vector space over $\mathbb{K}$. Some standard examples of modules are obtained in the following manner. Let I be an ideal of R and $\nu : R \to R/I$ the quotient map, then R/I is a module over R by defining $r \cdot \nu(m) := \nu(r \cdot m)$.

Definition 11.22. *Let M be a module over R. The annulator ideal is defined as*

$$\mathrm{Ann}(M) := \{\, r \in R \mid r \cdot m = 0, \ \forall m \in M \,\} \ .$$

That $\mathrm{Ann}(M)$ is an ideal is easy to check. It is also obvious that M is a module over $R/\mathrm{Ann}(M)$. In the above example by construction the ideal I is the annulator ideal of R/I. Hence every ideal of R is the annulator ideal of a suitable R-module.

Definition 11.23. *A module M is called a simple module if $M \neq \{0\}$ and M has only the trivial submodules $\{0\}$ and M.*

Proposition 11.24. *M is a simple module if and only if there is a maximal ideal P such that $M \cong R/P$.*

Proof. Note that the submodules of R/P correspond to the ideals lying between R and P. Hence, if P is maximal then R/P is simple. Conversely, given a simple module M take $m \in M, m \neq 0$. Then $R \cdot m$ is a submodule of M. Because $1 \cdot m = m$ we get for the module $R \cdot m \neq \{0\}$, hence it is the whole module M. The map $\varphi(r) = r \cdot m$ defines a surjective map $\varphi : R \rightarrow M$. This map is an R-module map where R is considered as a module over itself. The kernel P of such a map is an R-submodule. But R-submodules of R are nothing else than ideals of R. In view of the next section where we drop the commutativity let us note already that submodules of a ring R are more precisely the left ideals of R. The kernel P has to be maximal otherwise the image of a maximal ideal lying between P and R would be a nontrivial submodule of M. Hence $M \cong R/P$. $\square$

From this point of view the maximal ideals of $R(V)$ correspond to $R(V)$-module homomorphisms to simple $R(V)$-modules. If $R(V)$ is an algebra over the field $\mathbb{K}$, then a simple module M is of course a vector space over $\mathbb{K}$. By the above, we saw that it is even a field extension of $\mathbb{K}$. (Recall that $M \cong R/P$ with P a maximal ideal.) Because $R(V)$ is finitely generated as $\mathbb{K}$-algebra it is a finite-dimensional vector space over K (see [Ku, p. 56]), hence a finite (algebraic) field extension.

Proposition 11.25. *The maximal ideals (the "points") of $R = \mathbb{K}[X_1, X_2, \ldots, X_n]/I$ correspond to the $\mathbb{K}$-algebra homomorphism from R to arbitrary finite (algebraic) field extensions $\mathbb{L}$ of the base field $\mathbb{K}$. We call these homomorphisms $\mathbb{L}$-valued points.*

In particular, if the field $\mathbb{K}$ is algebraically closed there are no nontrivial algebraic field extensions. Hence there are only $\mathbb{K}$-valued points. If we consider reduced varieties (i.e. varieties whose coordinate rings are reduced rings) we get a complete dictionary. Let V be a variety, $P = I(V)$ the associated prime ideal generated as $P = (f_1, f_2, \ldots, f_r)$ with $f_i \in \mathbb{K}[X_1, X_2, \ldots, X_n]$, $R(V)$ the coordinate ring R_n/P, and $\bar{X}_i = X_i \bmod P$.
The points can be given in three ways:

(1) As classical points. $\alpha = (\alpha_1, \alpha_2, \ldots, \alpha_n) \in \mathbb{K}^n$ with
$f_1(\alpha) = f_2(\alpha) = \cdots = f_r(\alpha) = 0$.
(2) As maximal ideals in $R(V)$. They in turn can be identified with the maximal ideals in $\mathbb{K}[X_1, X_2, \ldots, X_n]$ which contain the prime ideal P. In an explicit manner these can be given as $(X_1 - \alpha_1, X_2 - \alpha_2, \ldots, X_n - \alpha_n)$ with the condition $f_1(\alpha) = f_2(\alpha) = \cdots = f_r(\alpha) = 0$.
(3) As surjective algebra homomorphisms $\phi : R(V) \rightarrow \mathbb{K}$. They are fixed by defining $\bar{X}_i \mapsto \phi(\bar{X}_i) = \alpha_i, i = 1, \ldots, n$ in such a way that
$\phi(f_1) = \phi(f_2) = \cdots = \phi(f_r) = 0$.

The situation is different if we drop the assumption that $\mathbb{K}$ is algebraically closed. The typical changes can already be seen if we take the real numbers $\mathbb{R}$ and the real affine line. The associated coordinate ring is $\mathbb{R}[X]$. There are only two finite extension fields of $\mathbb{R}$, either $\mathbb{R}$ itself or the complex number field $\mathbb{C}$. If we consider $\mathbb{R}$-algebra homomorphism from $\mathbb{R}[X]$ to $\mathbb{C}$ then they are given by prescribing $X \mapsto \alpha \in \mathbb{C}$. If $\alpha \in \mathbb{R}$ we are again in the same situation as above (this gives us the type (i) maximal ideals). If $\alpha \notin \mathbb{R}$ then the kernel I of the map is a maximal ideal of type (ii) with $I = (f)$ where f is a quadratic polynomial. f has α and $\bar{\alpha}$ as zeros. This says that the homomorphism $\Psi_{\bar{\alpha}} : X \mapsto \bar{\alpha}$ which is clearly different from $\Psi_{\alpha} : X \mapsto \alpha$ has the same kernel. In particular, for one maximal ideal of type (ii) we have two different homomorphisms. Note that the map $\alpha \to \bar{\alpha}$ is an element of the group of field automorphisms of $\mathbb{C}$ leaving $\mathbb{R}$ fixed, i.e. of the Galois group $G(\mathbb{C}/\mathbb{R}) = \{id, \tau\}$ where τ is complex conjugation.

The two homomorphisms Ψ_{α} and $\Psi_{\bar{\alpha}}$ are related as $\Psi_{\bar{\alpha}} = \tau \circ \Psi_{\alpha}$.

This is indeed the general situation for $R(V)$, a finitely generated $\mathbb{K}$-algebra. In general, there is no 1–1 correspondence between (1) and (2) anymore. But there is a 1–1 correspondence between maximal ideals of $R(V)$ and orbits of $\mathbb{K}$-algebra homomorphism of $R(V)$ onto finite field extensions $\mathbb{L}$ of $\mathbb{K}$ under the action of the Galois group

$$G(\mathbb{L}/\mathbb{K}) := \{ \sigma : \mathbb{L} \to \mathbb{L} \text{ an automorphism of fields with } \sigma_{|\mathbb{K}} = id \} .$$

11.4 Noncommutative Spaces

Up to now we had geometric objects and associated to them the dual objects, the algebra of functions on them. In our algebraic-geometric context these are the commutative coordinate rings (or coordinate algebras). If we consider the algebra of functions as the primary object and the space as the dual object and drop the commutativity for these algebras, we allow access to a completely new geometric world, the world of noncommutative spaces.

In this section I will only present the very basics of an algebraic-geometric approach to noncommutative spaces. Due to space limitations, differential geometric aspects (à la Connes) will be completely ignored.

For the following let R be a (not necessarily commutative) algebra over the field $\mathbb{K}$. First, we have to distinguish in this more general context left ideals (e.g. subrings I which are invariant under multiplication with R from the left), right ideals and two-sided ideals (which are left and right ideals). To construct quotient rings two-sided ideals are needed. If we use the term ideal without any additional comment we assume the ideal to be a two-sided one.

We want to introduce the concepts of prime ideals, maximal ideals, etc. A first definition of a prime ideal (starting directly from the definition in the commutative case) could be as follows. We call a two-sided ideal I *prime* if the quotient R/I contains no zero divisor. This definition has the drawback that there are rings without any prime ideal at all. Take for example the ring of 2×2 matrices. Besides the ideal $\{0\}$ and the whole ring the matrix ring does not contain any other ideal. To see this assume there is an ideal I

which contains a nonzero matrix A. By applying elementary operations from the left and the right we can transform any matrix to normal form which is a diagonal matrix with just 1 (at least one) and 0 on the diagonal. By multiplication with a permutation matrix we can achieve any pattern in the diagonal. These operations keep us inside the ideal. Adding suitable elements we see that the unit matrix is in the ideal. Hence the ideal is the whole ring. But obviously the matrix ring has zero divisors. Hence $\{0\}$ is not prime in this definition. We see that this ring does not contain any prime ideal at all with respect to the definition. We choose another name for such ideals: they are called *complete prime ideals*.

Definition 11.26. *A (two-sided) ideal I is called a prime ideal if for any two ideals J_1 and J_2 with $J_1 \cdot J_2 \subseteq I$ it follows that $J_1 \subseteq I$ or $J_2 \subseteq I$.*

This definition is equivalent to the following one.

Definition 11.27. *A (two-sided) ideal I is called a prime ideal if for any two elements $a, b \in R$ with $a \cdot R \cdot b \subseteq I$ it follows that $a \in I$ or $b \in I$.*

Proof. Def. (11.27) $\implies$ Def. (11.26): Let $J_1 \not\subseteq I$ and $J_2 \not\subseteq I$ ideals. We have to show that $J_1 \cdot J_2 \not\subseteq I$. For this choose $x \in J_1 \setminus I$ and $y \in J_2 \setminus I$. Then $x \cdot R \cdot y \subseteq J_1 \cdot J_2$ but there must be some $r \in R$ such that $x \cdot r \cdot y \notin I$ due to the condition that I is prime with respect to Def. (11.27). Hence $J_1 \cdot J_2 \not\subseteq I$ which is the claim.

Def. (11.26) $\implies$ Def. (11.27): Take $a, b \in R$. The ideals generated by these elements are RaR and RbR. The product of these "principal" ideals is not a principal ideal anymore. It is $RaR \cdot RbR = RaRbR := (arb \mid r \in R)$. Assume $arb \in I$ for all $r \in R$. Hence $(RaR)(RbR) \subseteq I$ and because I is prime we obtain by Def. (11.26) that either RaR or RbR as in I. Taking as element of R the 1 we get $a \in R$ or $b \in R$. $\square$

Every ideal which is a complete prime is prime. Obviously the condition in Def. 11.27 is a weaker condition than the condition that already from $a \cdot b \in I$ it follows that $a \in R$ or $b \in R$ (which is equivalent to: R/I contains no zero divisors). If R is commutative then they coincide. In this case $a \cdot r \cdot b = r \cdot a \cdot b$, and with $a \cdot b \in I$ also $r \cdot a \cdot b \in I$ which is no additional condition. Here you see clearly where the noncommutativity enters the picture. In the ring of matrices the ideal $\{0\}$ is prime because if after fixing two matrices A and B we obtain $A \cdot T \cdot B = 0$ for every T then either A or B has to be the zero matrix. This shows that the zero ideal in the matrix ring is a prime ideal.

   Maximal ideals are defined again as in the commutative setting just as maximal elements in the (nonempty) set of ideals. By Zorn's lemma there exist maximal ideals.

Proposition 11.28. *If M is a maximal ideal then it is a prime ideal.*

Proof. Take I and J ideals of R which are not contained in M. Then by the maximality of M we get $(I + M) = R$ and $(J + M) = R$, hence

$$R \cdot R = R = (I + M)(J + M) = I \cdot J + M \cdot J + I \cdot M + M \cdot M \ .$$

If we assume $I \cdot J \subseteq M$ then $R \subseteq M$ which is a contradiction. Hence $I \cdot J \not\subseteq M$. This shows M is prime. $\square$

By this result we see that every ring has prime ideals.

In the commutative case if we approach the theory of ideals from the point of view of modules over R we obtain an equivalent description. This is not true anymore in the noncommutative setting. For this let M be a (left-)module over R. As above we define

$$\text{Ann}(M) := \{\, r \in R \mid r \cdot m = 0, \forall m \in M \,\} \ ,$$

the annulator of the module M. $\text{Ann}(M)$ is a two-sided ideal. Clearly it is closed under addition and is a left ideal. (This is even true for an annulator of a single element $m \in M$). It is also a right ideal: let $s \in \text{Ann}(M)$ and $t \in R$ then $(st)m = s(tm) = 0$ because s annulates also tm.

Definition 11.29. *An ideal I is called a primitive ideal if I is the annulator ideal of a simple module M.*

Let us call the set of prime, resp. primitive, resp. maximal ideals $\text{Spec}(R)$, $\text{Priv}(R)$ and $\text{Max}(R)$.

Proposition 11.30.

$$\text{Spec}(R) \quad \supseteq \quad \text{Priv}(R) \quad \supseteq \quad \text{Max}(R) \ .$$

Proof. (1) Let P be a maximal ideal. Then R/P is a (left-)module. Unfortunately it is not necessarily simple (as module). The submodules correspond to left ideals lying between P and R. Choose Q a maximal left ideal lying above P. Then R/Q is a simple (left-)module and $P \cdot (R/Q) = 0$ because $P \cdot R = P \subseteq Q$. Hence $P \subseteq \text{Ann}(R/Q)$ and because $\text{Ann}(R/Q)$ is a two-sided ideal and P was a two-sided maximal ideal we get equality.
(2) Take $P = \text{Ann}(M)$, a primitive ideal. Assume P is not prime. Then there exist $a, b \in R$ but $a, b \notin P$ such that for all $r \in R$ we get $arb \in P$. This implies $arbm = 0$ for all $m \in M$ but $bm_0 \neq 0$ for at least one m_0. Now $B = R(bm_0)$ is a nonvanishing submodule. Obviously $a \in \text{Ann}(B)$, hence $B \neq M$. This contradicts the simplicity of M. $\square$

Clearly in the commutative case $\text{Priv}(R) = \text{Max}(R)$. Let me just give an example from [GOWA] that in the noncommutative case they fall apart. Take V an infinite-dimensional $\mathbb{C}$-vector space. Let R be the algebra of linear endomorphisms of V and I the nontrivial two-sided ideal consisting of linear endomorphisms with finite-dimensional image. The vector space V is an R-module by the natural action of the endomorphisms. We get that $V = R \cdot v$ where v is any nonzero vector of V. This implies that the module V is simple and that $\text{Ann}(V) = \{0\}$. Hence $\{0\}$ is primitive, but it is not maximal because I is lying above it.

In the commutative case we saw that we could interpret homomorphisms of the coordinate ring (which is an algebra if we consider varieties over a base field) into a field as points of the associated space. Indeed it is possible to give such an interpretation also in the noncommutative setting.

As a first example we consider the noncommutative affine plane. Recall that from the new point of view the "space" is the dual object to a possibly noncommutative algebra R. Very often spaces with quadratic relations are studied. Here we consider only the quantum plane (or Manin plane). First we consider the algebra of non-commutative polynomials in the variables X_1 and X_2. This is the free noncommutative algebra over the alphabet X_1 and X_2. Multiplication is defined by concatenation of the words. Now consider the two-sided ideal (i.e. the ideal of relations) generated by

$$X_1 X_2 = q X_2 X_1. \tag{11.15}$$

Depending on the situation one might consider q as another formal variable which commutes with everything, or it might be specialized to a certain number. In particular, by specializing $q = 1$ the coordinate functions commute and we obtain the usual commutative plane (which might be interpreted as the bosonic world). By specializing $q = -1$, the variables anticommute (and this might be interpreted as the fermionic world). In some sense one might interpret the quantum plane for a generic q as an interpolation (or passage) between the bosonic and the fermionic worlds.

The next example is an example of a quantum group. For details see [Magn]. Let $M_q(2)$ be the (noncommutative) $\mathbb{C}$-algebra generated by a, b, c, d, subject to the relations:

$$ab = \frac{1}{q} ba, \qquad ac = \frac{1}{q} ca, \qquad ad = da + \left(\frac{1}{q} - q\right) bc,$$
$$bc = cb, \qquad bd = \frac{1}{q} db, \qquad cd = \frac{1}{q} dc. \tag{11.16}$$

Again this is meant to take the free noncommutative algebra in the alphabet a, b, c, d and divide out the two-sided ideal generated by the expressions (left-side) $-$ (right-side) of all the relations (11.16) and build the quotient algebra. Here $q \in \mathbb{C}, q \neq 0$ (but an interpretation of q and $1/q$ as formal variables with the obvious relation is also possible).

Note that for $q = 1$ we obtain the commutative algebra of polynomial functions on the space of all 2×2 matrices over $\mathbb{C}$. In this sense the algebra $M_q(2)$ represents the "quantum matrices" as a "deformation of the usual matrices". To end up with the *quantum group* $Gl_q(2)$ we would have to add another element for the formal inverse of the quantum determinant $D = ad - (1/q)\, bc$.[1]

---

[1] There are other related objects which also carry the name quantum groups. In particular the approach via Hopf algebras is very important.

Now let A be another algebra. We call a $\mathbb{C}$-linear algebra homomorphism $\Psi \in \operatorname{Hom}(M_q(2), A)$ an *A-valued point* of $M_q(2)$. It is called a *generic point* if Ψ is injective. Saying that a linear map Ψ is an algebra homomorphism is equivalent to saying that the elements $\Psi(a), \Psi(b), \Psi(c), \Psi(d)$ fulfil the same relations (11.16) as the a, b, c and d. One might interpret Ψ as a point of the "quantum group". But be careful, it is only possible to "multiply" the two matrices if the images of the two maps

$$\Psi_1 \sim B_1 := \begin{pmatrix} a_1 & b_1 \\ c_1 & d_1 \end{pmatrix}, \qquad \Psi_2 \sim B_2 := \begin{pmatrix} a_2 & b_2 \\ c_2 & d_2 \end{pmatrix}$$

lie in a common algebra A_3, i.e. $a_1, b_1, c_1, d_1 \in A_1 \subseteq A_3$ and $a_2, b_2, c_2, d_2 \in A_2 \subseteq A_3$. Then we can multiply the two matrices $B_1 \cdot B_2$ as prescribed by the usual matrix product and obtain another matrix B_3 with coefficients $a_3, b_3, c_3, d_3 \in A_3$. This matrix defines only a homomorphism of $M_q(2)$, i.e. an A_3-valued point, if $\Psi_1(M_q(2))$ commutes with $\Psi_2(M_q(2))$ as subalgebras of A_3. In particular, the product of Ψ with itself is not an A-valued point of $M_q(2)$ anymore. One can show that it is an A-valued point of $M_{q^2}(2)$.

Hints for Further Reading

For the necessary background on commutative algebras see the book of Kunz [Ku]. It has the advantage that it stays always in close connection to the geometric interpretation of the algebraic results.

For the noncommutative situation (in relation to physical interpretations) further details can be found in article by Julius Wess and Bruno Zumino [WZ], where one finds references for further study in this direction. Quantum planes and noncommutative algebraic geometry can be found, e.g. in some articles and lectures by Manin "Quantum Groups and Noncommutative Geometry" [Magn], "Topics in Noncommutative Geometry" [Mang] and "Notes on Quantum Groups and the Quantum de Rham Complexes" [Maqg] and Artin "Geometry of Quantum Planes" [Arq]. See also Rosenberg [Ros].

For the general noncommutative situation I like to recommend Goodearl and Warfield, "An Introduction to Noncommutative Noetherian Rings" [GOWA] and Borho, Gabriel, Rentschler, "Primideale in Einhüllenden auflösbarer Liealgebren" [BGR]. These books are completely on the algebraic side of the theory.

[Arq]   Artin, M., *Geometry of Quantum Planes* (in) *Azumaya Algebras, Actions, and Modules, Proceedings Bloomington 1990*, Contemp. Math. **124**, 1–15, AMS, Providence.

[BGR]   Borho, W., Gabriel, P., Rentschler, R., *Primideale in Einhüllenden auflösbarer Lie-Algebren*, Lecture Notes in Mathematics **357**, Springer, 1973.

[GOWA] Goodearl, K.R., Warfield, R.B., *An Introduction to Noncommutative Noetherian Rings*, London Math. Soc. Student Texts **16**, Cambridge University Press, New York, 1989.

[Ku]   Kunz, E., *Einführung in die kommutative Algebra und algebraische Geometrie*, Vieweg, Braunschweig, 1979; (in English translation) *Introduction to Commutative Algebra and Algebraic Geometry*, Birkhäuser, Boston, 1985.

[Maqg] Manin, Y.I., *Notes on Quantum Groups and Quantum de Rham Complexes*, MPI/91-60.

[Mang] Manin, Y.I., *Topics in Noncommutative Geometry*, M. B. Porter Lectures, Princeton University Press, Princeton, New Jersey, 1991.

[Magn] Manin, Y.I., *Quantum Groups and Noncommutative Geometry*, Université de Montréal (CRM), 1988.

[Ros]   Rosenberg, Alexander, L., The Left Spectrum, the Levitzki Radical, and Noncommutative Schemes, *Proc. Natl. Acad. Sci. USA* **87**(4) (1990) 8583–8586.

[WZ]   Wess, J., Zumino, B., Covariant Differential Calculus on the Quantum Hyperplane, *Nucl. Phys. B Proceed.* Suppl. **18B** (1990) 302.

12

Schemes

Differentiable manifolds look locally like the real number space $\mathbb{R}^n$ with its standard differentiable structure. In other words, $\mathbb{R}^n$ is our "local model" for differentiable manifolds and every differentiable manifold looks locally like the unique model manifold. The unique local model of complex manifolds is $\mathbb{C}^n$ with its standard complex structure. Affine schemes are the local "model spaces" of algebraic geometry and general schemes will look locally like affine schemes. Contrary to the differentiable setting, there is not just one model space but a lot of them.

12.1 Affine Schemes

For this section let R be again a commutative ring with unit 1. We do not assume R to be an algebra over a field $\mathbb{K}$.

Affine schemes are very useful generalizations of affine varieties. Starting from affine varieties V over a field $\mathbb{K}$ we saw that we were able to assign dual objects to them, the coordinate rings $R(V)$. The geometric structure of V (subvarieties, points, maps, ...) are represented by the algebraic structure of $R(V)$ (prime ideals, maximal ideals, ring homomorphisms, ...). After dualization we are even able to extend our notion of "space" in the sense that we can consider more general rings and regard them as dual objects of some generalized "spaces". In noncommutative quantum geometry one even studies certain noncommutative algebras over a field $\mathbb{K}$. Quantum spaces are the dual objects of these algebras. In the following we will restrict ourselves to the commutative case, but we will allow arbitrary rings.

What are the dual objects (dual to the rings) which generalize the concept of a variety? We saw already that prime ideals of the coordinate ring correspond to subvarieties and that Zariski closed prime ideals (at least if the field $\mathbb{K}$ is algebraically closed) correspond to points. It is quite natural to take as space the set $\mathrm{Spec}(R)$ together with its Zariski topology. But this is not enough. If we take for example $R_1 = \mathbb{K}$ and $R_2 = \mathbb{K}[\epsilon]/(\epsilon^2)$ then in both cases

M. Schlichenmaier, Schemes. In: M. Schlichenmaier, An Introduction to Riemann Surfaces,
Algebraic Curves and Moduli Spaces, Theoretical and Mathematical Physics, 155–168 (2007)
DOI 10.1007/978-3-540-71175-9_13

$\mathrm{Spec}(R_i)$ consists just of one point. It is represented in the first case by the ideal $\{0\}$ and in the second case by (ϵ). Obviously both Spec coincide. Let us compare this with the differentiable setting. For an arbitrary differentiable manifold the structure is not yet given if we consider the manifold just as a topological manifold. We have to fix its differentiable structure. This is done if we tell what the differentiable functions are.

The same is necessary in the algebraic situation. Hence $\mathrm{Spec}(R)$ together with the functions (which in the case of varieties correspond to the elements of R) should be considered as "space". So the space associated to a ring R should be $(\mathrm{Spec}(R), R)$. In fact, $\mathrm{Spec}(R)$ is not a data independent of R. Nevertheless we will write both information in view of globalizations of the notion. Compare this again with the differentiable situation. If you have a manifold which is $\mathbb{R}^n$ (the model manifold) then the topology is fixed. But if you have an arbitrary differentiable manifold then you need a topology in the first place to define coordinate charts. In view of these globalizations we additionally have to replace the ring of functions by a data which will give us all local and global functions together. Note that in the case of compact complex analytic manifolds there would exist no nonconstant global analytic functions.

The right setting for this is the language of sheaves. Sheaves of abelian groups have been introduced in Sect. 8.2. Sheaves of rings are introduced in a completely analogous way. For an exact definition of a sheaf of rings consult the second paragraph of Sect. 8.2 and replace "abelian group" by "commutative ring" and "group homomorphism" by ring homomorphism. A sheaf is the coding of an object which is local and global in a compatible way. A standard example (which is in some sense too simple) is the sheaf of differentiable functions on a differentiable manifold X. It assigns to every open set U the ring of differentiable functions defined on U. The compatibility just means that this assignment is compatible with the restriction of the sets where the functions are defined on.

Given a ring R its associated affine scheme is the pair $(\mathrm{Spec}(R), \mathcal{O}_R)$ where $\mathrm{Spec}(R)$ is the set of prime ideals made into a topological space by the Zariski topology and $\mathcal{O}_R$ is a sheaf of rings on $\mathrm{Spec}(R)$ which we will define in a minute. The sheaf $\mathcal{O}_R$ is called the structure sheaf. For simplicity this pair is sometimes just called $\mathrm{Spec}(R)$.

Recall that the sets $V(S) := \{P \in \mathrm{Spec}(R) \mid P \supseteq S\}$, where S is any subset of R, are the closed sets of the Zariski topology of $X := \mathrm{Spec}(R)$. Hence the sets $\mathrm{Spec}(R) \backslash V(S)$ are exactly its open sets. There are the following special open sets in X. For a single element $f \in R$ we define

$$X_f := \mathrm{Spec}(R) \setminus V(f) = \{ P \in \mathrm{Spec}(R) \mid f \notin P \} . \tag{12.1}$$

The set $\{X_f, f \in R\}$ is a basis of the topology which says that every open set is a union of such X_f, see Lemma 12.9. This is especially useful because the X_f are again affine schemes. More precisely, $X_f = \mathrm{Spec}(R_f)$. Here the ring R_f is defined as the ring of fractions with the powers of f as denominators:

$$R_f := \left\{ \frac{g}{f^n} \mid g \in R, \ n \in \mathbb{N}_0 \right\} .$$

Let me explain this construction. It is a generalization of the way how one constructs the rational numbers from the integers. Let S be a multiplicative system of R, i.e. a subset of R which is multiplicatively closed and contains 1. (In our example, $S := \{1, f, f^2, f^3, \ldots\}$.) Now introduce on the set of pairs $R \times S$ the equivalence relation

$$(t, s) \sim (t', s') \iff \exists s'' \in S \text{ such that } s''(s't - st') = 0 .$$

The equivalence class of (s, t) is denoted by s/t. There is always a map $R \to R_f$ given by $r \mapsto r/1$. The ideals in R_f are obtained by mapping the ideals I of R to R_f and multiplying them by $R_f : R_f \cdot I$. By construction, f is a unit in R_f. Hence if $f \in P$ where P is any ideal then $R_f = R_f \cdot P$. If P is a prime ideal of R such that $f \notin P$ then $R_f \cdot P$ is a prime ideal of R_f. This shows $X_f = \operatorname{Spec}(R_f)$. For details see [Ku].

If f is not a zero divisor the map $R \to R_f$ is an embedding and if f is not a unit in R the ring R_f will be bigger. This is completely in accordance with our understanding of R, resp. R_f, as functions on X, resp. on the honest subset X_f. Passing from X to X_f is something like passing from the global to the more local situation. This explains why this process of taking the ring of fractions with respect to some multiplicative subset S is sometimes called localization of the ring.

Now we define our sheaf $\mathcal{O}_R$ for the basis sets X_f. In $X_f \cap X_g$ are the prime ideals which neither contain f nor g. Hence they do not contain $f \cdot g$. It follows that $X_f \cap X_g = X_{fg}$. We see that the set of the X_f are closed under intersections. Note also that $X_1 = X$ and $X_0 = \emptyset$. We define

$$\mathcal{O}_R(X) := R, \qquad \mathcal{O}_R(X_f) := R_f . \tag{12.2}$$

For $X_{fg} = X_f \cap X_g \subseteq X_f$ we define the restriction map

$$\rho_{fg}^f : R_f \to (R_f)_g = R_{fg}, \qquad r \mapsto \frac{r}{1} .$$

It is easy to check that all the maps $\rho_{\cdot\cdot}^{\cdot}$ are compatible on the intersections of the basis open sets. In Sect. 12.3 we will show that the other sheaf axioms are fulfilled for the X_f with respect to their intersections. Hence we have defined the sheaf $\mathcal{O}_R$ on the basis of the topology which is closed under intersections. The whole sheaf is now defined by some general construction. We set

$$\mathcal{O}_R(U) := \operatorname*{proj\,lim}_{X_f \subseteq U} \mathcal{O}_R(X_f)$$

for U a general open set. For more details see [EHS]. Let us collect the facts.

Definition 12.1. *Let R be a commutative ring. The pair $(\mathrm{Spec}(R), \mathcal{O}_R)$, where $\mathrm{Spec}(R)$ is the space of prime ideals with the Zariski topology and $\mathcal{O}_R$ is the sheaf of rings on $\mathrm{Spec}(R)$ introduced above, is called the associated affine scheme $\mathrm{Spec}(R)$ of R. The sheaf $\mathcal{O}_R$ is called the structure sheaf of $\mathrm{Spec}(R)$.*

Let me explain in which sense the elements f of an arbitrary ring R can be considered as functions, i.e. as prescriptions how to assign a value from a field to every point. This gives me the opportunity to introduce another important concept which is related to points: the residue fields. Fix an element $f \in R$. Let $[P] \in \mathrm{Spec}(R)$ be a (not necessarily closed) point, i.e. P is a prime ideal. We define

$$f([P]) := f \mod P \in R/P \tag{12.3}$$

in a first step. From the primeness of P it follows that R/P is an integral domain ring (i.e. it contains no zero divisor). Hence $S := (R/P) \setminus \{0\}$ is a multiplicative system and the ring of fractions, denoted by $\mathrm{Quot}(R/P)$, is a field, the quotient field. Because R/P is an integral domain it can be embedded into its quotient field. Hence $f([P])$ is indeed an element of a field. Contrary to the classical situation, if we change the point $[P]$ the field $\mathrm{Quot}(R/P)$ will change too.

Example 12.2. Take again $R = \mathbb{C}[X, Y]$ and $f \in R$. Here we have three different types of points in $\mathrm{Spec}(R)$.
Type (i): the closed points $[M]$ with $M = (X - \alpha, Y - \beta)$ a maximal ideal. We write $f = f_0 + (X - \alpha) \cdot g + (Y - \beta) \cdot h$ with $f_0 = f(\alpha, \beta) \in \mathbb{C}$ and $g, h \in R$. Now

$$f([M]) = f \mod M = f_0 + (X - \alpha) \cdot g + (Y - \beta) \cdot h \mod M = f_0 \ .$$

The quotient R/M is already a field, hence it is the residue field. In our case it is always the base field $\mathbb{C}$ independent of the point $[M]$. The value $f([M])$ is just the value we obtain by plugging the point (α, β) into the polynomial f. Note that the points are subvarieties of dimension 0.
Type (ii): the points $[P]$ with $P = (h)$, a principal ideal. Here h is an irreducible polynomial in the variables X and Y. If we calculate R/P we obtain $\mathbb{C}[X, Y]/(h)$ which is not a field. As residue field we obtain $\mathbb{C}(X, Y)/(h)$. This field consists of all rational expressions in the variables X and Y with the relation $h(X, Y) = 0$. This implies that the transcendence degree of the residue field over the base field is one, i.e. one of the variables X or Y is algebraically independent over $\mathbb{C}$ and the second variable is in an algebraic relation with the first and the elements of $\mathbb{C}$. Note that the coordinate ring has (Krull-) dimension one and the subvariety corresponding to $[P]$ is a curve, i.e. is an object of geometric dimension one.
Type (iii): $[\{0\}]$ the zero ideal. In this case $R/P = \mathbb{C}[X, Y]$ and the residue field is $\mathbb{C}(X, Y)$ the rational function field in two variables. In particular, its transcendence degree is two and coincides with the (Krull-)dimensions of

the coordinate ring and the geometric dimension of the variety $V(\{0\})$ which equals the whole affine plane $\mathbb{C}^2$.

Strictly speaking, we have not shown (and will not do it here) that there are no other prime ideals. But this is in fact true, see [Ku]. The equality of the transcendence degree of the residue field and the (Krull-)dimension of the coordinate ring obtained above is true for all varieties over arbitrary fields. For example, if we replace $\mathbb{C}$ by $\mathbb{R}$ we obtain for the closed points, the maximal ideals, either $\mathbb{R}$ or $\mathbb{C}$ as residue fields. Both fields have transcendence degree 0 over $\mathbb{R}$.

Example 12.3. Consider $R = \mathbb{Z}$, the integers, then $\mathrm{Spec}(\mathbb{Z})$ consists of the zero ideal and the principal ideals generated by prime numbers. As residue field we obtain for $[\{0\}]$ the field $\mathrm{Quot}(\mathbb{Z}/(0)) = \mathbb{Q}$ and for the point $[(p)]$ (which is a closed point) $\mathbb{F}_p = \mathbb{Z}/(p)$, the prime field of characteristic p. In particular, we see at this example that even for the maximal points the residue field can vary in an essential way. Note that $\mathbb{Z}$ is not an algebra over a fixed base field.

Up to now we considered one ring, resp. one scheme. In any category of objects one has maps between the objects. Let $\Phi : R \to S$ be a ring homomorphism. If I is any ideal of S, then $\Phi^{-1}(I)$ is an ideal of R. The reader is advised to check that if P is prime then $\Phi^{-1}(P)$ is again prime. Hence $\Phi^* : P \mapsto \Phi^{-1}(P)$ is a well-defined map $\mathrm{Spec}(S) \to \mathrm{Spec}(R)$. Indeed, it is even continuous because the pre-image of a closed set is again closed. Let $X = (\mathrm{Spec}(S), \mathcal{O}_S)$ and $Y = (\mathrm{Spec}(R), \mathcal{O}_R)$ be two affine schemes. The map Φ also induces a map on the level of the structure sheaves $\Phi_* : \mathcal{O}_R \to \mathcal{O}_S$. The pair (Φ^*, Φ_*) of maps fulfils certain compatibility conditions which makes them to a homomorphism of schemes, see Definition 12.5.

12.2 General Schemes

A already said above, schemes are "spaces" which are locally affine schemes. To be more precise

Definition 12.4. *(a) A scheme is a pair $X = (|X|, \mathcal{O}_X)$ consisting of a topological space $|X|$ and a sheaf $\mathcal{O}_X$ of rings on X, such that X is locally isomorphic to affine schemes $(\mathrm{Spec}(R), \mathcal{O}_R)$ associated to rings R. This says that for every point $x \in X$ there is an open set U containing x, and a ring R (it may depend on the point x) such that the affine scheme $(\mathrm{Spec}(R), \mathcal{O}_R)$ associated to R is isomorphic to the scheme $(U, \mathcal{O}_{X|U})$. In other words there is a homeomorphism $\Psi : U \to \mathrm{Spec}(R)$ such that there is an isomorphism of sheaves*

$$\Psi^\# : \mathcal{O}_R \cong \Psi_*(\mathcal{O}_{X|U}) . \tag{12.4}$$

Here the sheaf $\Psi_*(\mathcal{O}_{X|U})$ is defined to be the sheaf on $\mathrm{Spec}(R)$ given by the assignment

$$\Psi_*(\mathcal{O}_{X|U})(W) := \mathcal{O}_X(\Psi^{-1}(W)), \qquad \text{for every open set } W \subseteq \operatorname{Spec}(R).$$

(b) A scheme is called an affine scheme if it is globally isomorphic to an affine scheme $(\operatorname{Spec}(R), \mathcal{O}_R)$ *associated to a ring* R.

Definition 12.5. *Given two schemes* $X = (|X|, \mathcal{O}_X)$ *and* $Y = (|Y|, \mathcal{O}_Y)$. *A morphism of schemes is a pair of maps*

$$(\psi, \psi^{\#}) : X = (|X|, \mathcal{O}_X) \quad \rightarrow \quad Y = (|Y|, \mathcal{O}_Y) \tag{12.5}$$

such that $\psi : |X| \rightarrow |Y|$ *is a continuous map and* $\psi^{\#} : \mathcal{O}_Y \rightarrow \psi_*\mathcal{O}_X$ *is a homomorphism of sheaves of rings which is local.*
Such a homomorphism of sheaves is called local if the following is true. Take every point $p \in X$ *and* $q = \psi(p)$ *its image point in* Y. *Take every* $f \in \mathcal{O}_Y(U)$ *then* $\psi^{\#}f \in \mathcal{O}_X(\psi^{-1}(U))$. *The map* $\psi^{\#}$ *is local if* $f(p) = 0$ *if and only if* $\psi^{\#}f(q) = 0$ *(for every* p *and* f). *Here we take the interpretation of the elements* f *and* $\psi^{\#}f$ *as functions following (12.3).*

Remark 12.6. There is another way to define locality. It is possible to introduce for the structure sheaf of a scheme the stalk at a given point, i.e. $\mathcal{O}_{X,p}$, which is a local ring, i.e. it has only one maximal ideal. The maximal ideal is given by the elements vanishing (in the interpretation as functions (12.3)) at the point. Denote this ideal by $M_{X,p}$. Then $\psi^{\#}$ is local if for every $p \in X$ the induced map fulfils $\psi^{\#}_{\psi(p)} : M_{Y,\psi(p)} \rightarrow M_{X,p}$, i.e. it is a local map in the terminology of local rings.

Remark 12.7. As for schemes the stalks are local rings, schemes are examples of *locally ringed spaces*. Morphisms between schemes should be morphism of locally ringed spaces. Their definition corresponds exactly to the requirement that $\psi^{\#}$ is local.

Proposition 12.8. *The category of affine schemes is equivalent to the category of commutative rings with unit with the arrows (representing the maps) reversed.*

There are other important concepts in this theory. First, there is the concept of a scheme over another scheme. This is the right context to describe families of schemes. Only within this framework it is possible to make such useful things precise as degenerations, moduli spaces, etc. Note that every affine scheme is in a natural way a scheme over $\operatorname{Spec}(\mathbb{Z})$, because for every ring R we have the natural map

$$\mathbb{Z} \rightarrow R, \quad n \mapsto n \cdot 1.$$

Taking the dual map introduced above we obtain a homomorphism of schemes.

If R is a $\mathbb{K}$-algebra with $\mathbb{K}$ a field then we have the map $\mathbb{K} \to R$, $\alpha \mapsto \alpha \cdot 1$, which is a ring homomorphism. Hence we always obtain a map:

$$\mathrm{Spec}(R) \to \mathrm{Spec}(\mathbb{K}) = (\{0\}, K) \ .$$

By considering the coordinate ring $R(V)$ of an affine variety V over a fixed algebraically closed field $\mathbb{K}$ and assigning to it the affine scheme $\mathrm{Spec}(R(V))$ we obtain a map (i.e. a functor) from the category[1] of varieties over $\mathbb{K}$ to the category of schemes over $\mathbb{K}$. The schemes corresponding to varieties are the *irreducible and reduced noetherian affine schemes of finite type over* $\mathrm{Spec}(\mathbb{K})$. The additional properties of the scheme are nothing else as the corresponding properties for the defining ring $R(V)$. Here finite type means that $R(V)$ is a finitely generated $\mathbb{K}$-algebra. One sees again in which sense the schemes extend our geometric objects from the varieties to more general "spaces".

The second concept is the concept of a *functor of points* of a scheme. We saw already at several places that points of a geometric object can be described as homomorphisms of the dual (algebraic) object into some simple (algebraic) object. Fix X a scheme and take for S any scheme and denote by $\mathrm{Hom}(S, X)$ the set of scheme morphisms from S to X. Any such a morphism is called an *S-valued point* of X. Note that we are in the geometric category, hence the order of the elements in $\mathrm{Hom}(.,.)$ is just the other way around compared to the former section. If we vary S (but keep X fixed) we get a map h_X assigning to every scheme S the set $h_X(S) = \mathrm{Hom}(S, X)$. This assignment h_X is functorial, in the sense that it also makes a compatible assignment of morphism between schemes $S \to T$ to maps between the sets $h_X(T) \to h_X(S)$ by composition. In the language of category theory h_X is a (contravariant) functor. The functor h_X is called the *functor of points* associated to X. Now X is completely fixed by the functor h_X. In categorical language: X represents its own functor of points. The advantage of this view-point is that certain questions of algebraic geometry, like the existence of a moduli space for certain geometric data, can be easily transferred to the language of functors. One can extract a lot of geometric data without knowing whether there is indeed a scheme having this functor as functor of points (i.e. having a scheme representing the functor). In the case that such a scheme exists, the functor is called representable. If you want to know more about this beautiful subject you should consult [EHS] and [Mpg].

---

[1] For the reader not acquainted with the categorical language: Roughly speaking a category is a collection of objects with maps (morphisms, arrows) between them. A functor is a "map" between two categories, mapping objects to objects and morphism (arrows) to morphism (arrows) in a compatible way. A covariant functor respects the direction of the arrow, a contravariant functor inverts the direction.

12.3 The Structure Sheaf $\mathcal{O}_R$

In this sect. I would like to show that the sheaf axioms for the structure sheaf $\mathcal{O}_R$ on $X = \operatorname{Spec}(R)$ are fulfilled if we consider only the basis open sets $X_f = \operatorname{Spec}(R) \setminus V(f)$. As the understanding of this construction is not really important for the following the reader might skip it for the first reading.

Recall that the intersection of two basis open sets $X_f \cap X_g = X_{fg}$ is again a basis open set. The sheaf $\mathcal{O}_R$ on the basis open sets was defined to be $\mathcal{O}_R(X_f) := R_f$ and the restriction maps were the natural maps

$$R_f \;\to\; (R_f)_g \;=\; R_{fg}, \qquad r \mapsto \frac{r}{1}\,.$$

Lemma 12.9. *The set* $\{X_f \mid f \in R\}$ *is a basis of the topology.*

Proof. We have to show that every open set U is a union of such X_f. By definition, there is a $S \subseteq R$ such that

$$U = \operatorname{Spec}(R)\setminus V(S) = \operatorname{Spec}(R)\setminus(\bigcap_{f \in S} V(f)) = \bigcup_{f \in S}(\operatorname{Spec}(R)\setminus V(f)) = \bigcup_{f \in S} X_f\,.\quad\square$$

Obviously only a set of generators $\{f_i \mid i \in J\}$ of the ideal generated by the set S is needed. Hence if R is a noetherian ring every open set can already be covered by finitely many X_f.

Lemma 12.10. *Let* $X = \operatorname{Spec}(R)$ *and* $\{f_i\}_{i \in J}$ *a set of elements of R then the union of the sets X_{f_i} equals X if and only if the ideal generated by the f_i equals the whole ring R.*

Proof. The union of the X_{f_i} covers $\operatorname{Spec}(R)$ if and only if no prime ideal of R contains all the f_i. But every ideal strictly smaller than the whole ring is dominated by a maximal (and hence prime) ideal. Hence $\operatorname{Spec} R$ is covered by them if and only if the ideal generated by the f_i is the whole ring. $\square$

Lemma 12.11. *The affine scheme* $X = \operatorname{Spec}(R)$ *is a quasi-compact space. This says every open cover of X has a finite subcover.*

Proof. Let $X = \bigcup_{j \in J} X_j$ be a cover of X. Because the basis open set X_f are a basis of the topology, every X_j can be given as union of X_{f_i}. Altogether we get a refinement of the cover $X = \bigcup_{i \in I} X_{f_i}$. By Lemma 12.10 the ideal generated by these f_i is the whole ring. In particular, 1 is a finite linear combination of the f_i. Taking only these f_i which occur with a nonzero coefficient in the linear combination we get (using Lemma 12.10 again) that $X_{f_{i_k}}$, $k = 1, .., r$ is a finite subcover of X. Taking for every k just one element X_{j_k} containing $X_{f_{i_k}}$ we obtain a finite number of sets which is a subcover from the cover we started with. $\square$

Note that this space is not called a compact space because the Hausdorff condition that every distinct two points have disjoint open neighbourhoods is obviously not fulfilled.

The following proposition says that the sheaf conditions (3) and (4) from Sect. 8.2 for the basis open sets are fulfilled.

Proposition 12.12. *Let X_f be covered by $\{X_{f_i}\}_{i \in I}$.*

(a) Let $g, h \in R_f = \mathcal{O}_R(X_f)$ with $g = h$ as elements in $R_{f_i} = \mathcal{O}_R(X_{f_i})$ for every $i \in I$, then $g = h$ also in R_f.

(b) Let $g_i \in R_{f_i}$ be given for all $i \in I$ with $g_i = g_j$ in $R_{f_i f_j}$, then there exists a $g \in R_f$ with $g = g_i$ in R_{f_i}.

Proof. Because $X_f = \mathrm{Spec}(R_f)$ is again an affine scheme it is enough to show the proposition for $R_f = R$, where R is an arbitrary ring. Let $\mathrm{Spec}\, R = X = \bigcup_{i \in I} X_{f_i}$.

(a) Let $g, h \in R$ be such that they map to the same element in R_{f_i}. This can only be the case if in R we have

$$f_i^{n_i} \cdot (g - h) = 0, \qquad \forall i \in I,$$

(see the construction of the ring of fractions above). Due to the quasi-compactness it is enough to consider finitely many f_i, $i = 1, .., r$. Hence there is a N such that for every i the element f_i^N annulates $(g - h)$. There is another number M, depending on N and r, such that we have the following inclusion of ideals

$$(f_1^N, f_2^N, \ldots, f_r^N) \supseteq (f_1, f_2, \ldots, f_r)^M .$$

Because the X_{f_i}, $i = 1, .., r$, are a cover of X the ideal on the right side equals always the ideal (1), see Lemma 12.10. Hence, also the ideal on the left. Combining 1 as linear combination of the generator we get

$$1 \cdot (g - h) = (c_1 f_1^N + c_2 f_2^N \cdots + c_r f_r^N)(g - h) = 0 .$$

This shows (a).

(b) Let $g_i \in R_{f_i}$, $i \in I$ be given such that $g_i = g_j$ in $R_{f_i f_j}$. This says there exists a N such that

$$(f_i f_j)^N g_i = (f_i f_j)^N g_j$$

in R. Note that every g_i can be written as $g_i^* / f_i^{k_i}$ with $g_i^* \in R$. Hence if N is big enough the elements $f_i^N g_i$ are in R. Again by the quasi-compactness a common N will do it for every pair (i, j). Using the same arguments as in (a) we get

$$1 = \sum_i e_i f_i^N, \qquad e_i \in R .$$

This formula corresponds to a "partition of unity". We set

$$g = \sum e_i f_i^N g_i .$$

We get

$$f_j^N g = \sum_i f_j^N e_i f_i^N g_i = \sum_i e_i f_i^N f_j^N g_j = f_j^N g_j .$$

This shows $g = g_j$ in R_{f_j}. $\quad\square$

12.4 Examples of Schemes

12.4.1 Projective Varieties

Affine varieties are examples of affine schemes over a field $\mathbb{K}$. Projective varieties are schemes. Recall from Chap. 6 the definition of the *projective space* $\mathbb{P}^n(\mathbb{K})$ of dimension n over a field $\mathbb{K}$. There we considered only $\mathbb{K} = \mathbb{C}$. But the same construction works for arbitrary fields $\mathbb{K}$. The projective space can be given as orbit space $(\mathbb{K}^{n+1} \setminus \{0\})/ \sim$, where two $(n + 1)$-tuple α and β are equivalent if $\alpha = \lambda \cdot \beta$ with $\lambda \in \mathbb{K}$, $\lambda \neq 0$. *Projective varieties* are defined to be the vanishing sets of homogeneous polynomials in $n + 1$ variables, see Chap. 6.

Again everything can be dualized. One considers the projective coordinate ring and its set of homogeneous ideals (ideals which are generated by homogeneous elements). In the case of $\mathbb{P}^n(\mathbb{K})$ the homogeneous coordinate ring is $\mathbb{K}[Y_0, Y_1, \ldots, Y_n]$. Again it is possible to introduce the Zariski topology on the set of homogeneous prime ideals. It is even possible to introduce the notion of a projective scheme Proj, which is again a topological space together with a sheaf of rings, see [EHS].

In the same way as $\mathbb{P}^n(\mathbb{K})$ can be covered by $(n + 1)$ affine spaces $\mathbb{K}^n$ it is possible to cover every projective scheme by finitely many affine schemes. This covering is even such that the projective scheme is locally isomorphic to these affine schemes. Hence it is a scheme. The projective scheme $\mathrm{Proj}(\mathbb{K}[Y_0, Y_1, \ldots, Y_n])$ is locally isomorphic to $\mathrm{Spec}(\mathbb{K}[X_1, X_2, \ldots, X_n])$. For example, the open set of elements α with $Y_0(\alpha) \neq 0$ is in 1–1 correspondence to it via the assignment $X_i \mapsto Y_i/Y_0$.

As already said, the projective schemes are schemes and you might ask why should one pay special attention to them. Projective schemes are quite useful. They are schemes with rather strong additional properties. For example, in the classical case (e.g. nonsingular varieties over $\mathbb{C}$) projective varieties are compact in the classical complex topology. This yields all the interesting results like: there are no nonconstant global analytic or harmonic functions, the theorem of Riemann–Roch is valid, the integration is well defined and so on. Indeed, we get similar results for projective schemes over arbitrary fields. Here it is the feature "properness" which generalizes compactness.

12.4.2 The Scheme of Integers

The affine scheme $\mathrm{Spec}(\mathbb{Z}) = (\mathrm{Spec}(\mathbb{Z}), \mathcal{O}_{\mathrm{Spec}(\mathbb{Z})})$ which we discussed earlier. The topological space consists of the element $[\{0\}]$ and the elements $[(p)]$ where p takes every prime number. The residue fields are $\mathbb{Q}$, resp. the finite fields $\mathbb{F}_p$. What are closed sets. By definition, these are exactly the sets $V(S)$ such that there is a $S \subseteq \mathbb{Z}$ with

$$V(S) := \{\, [(p)] \in \mathrm{Spec}(\mathbb{Z}) \mid (p) \supseteq S \,\} = V((S)) = V((gcd(S)))\,.$$

Here gcd denotes the greatest common divisor. For the last identification recall that the ideal (S) has to be generated by one element n because $\mathbb{Z}$ is a principal ideal ring. Now every element in S has to be a multiple of this n. We have to take the biggest such n which fulfils this condition, hence $n = gcd(S)$. If $n = 0$ then $V(n) = V(0) = \mathrm{Spec}(\mathbb{Z})$, if $n = 1$ then $V(n) = V(1) = \emptyset$, otherwise $V(n)$ consists of the finitely many primes, resp. their ideals, dividing n. Altogether we get that the closed sets are, besides the whole space and the empty set, just sets of finitely many points. As already said at some other place $\mathbb{Z}$ resembles $\mathbb{K}[X]$ very much. By the way, we see that the topological closure $\overline{[\{0\}]} = \mathrm{Spec}(\mathbb{Z})$ is the whole space. For this reason $[\{0\}]$ is called the generic point of $\mathrm{Spec}(\mathbb{Z})$.

All these have important consequences. We have two principles which can be very useful:

(1) Let some property be defined over $\mathbb{Z}$ and assume it is a closed property. Assume further that the property is true for infinitely many primes (e.g. the property is true if we consider the problem in characteristic p for infinitely many p) then it has to be true for the whole $\mathrm{Spec}(\mathbb{Z})$. Especially it has to be true for all primes and for the generic point, i.e. in characteristic zero.

(2) Now assume that the property is an open property. If it is true for at least one point, then it is true for all points except for possibly finitely many points. In particular, it has to be true for the generic point (characteristic zero) because every nonempty open set has to contain the generic point.

12.4.3 A Family of Curves

This example illustrates the second principle above. I took the example from [EHS]. You are encouraged to develop your own examples. Consider the conic

$$X^2 - Y^2 = 5.$$

It defines a curve in the real (or complex) plane. In fact, it is already defined over the integers which says that there is a defining equation for the curve with integer coefficients. Hence it makes perfect sense to ask for points $(\alpha, \beta) \in \mathbb{Z}^2$ which solve the equation. We already saw that it is advantageous to consider the coordinate ring. The coordinate ring and everything else also make sense if there would be no integer solution at all. Here we have

$$\mathbb{Z} \rightarrow R = \mathbb{Z}[X,Y]/(X^2 - Y^2 - 5), \qquad \mathrm{Spec}(R) \rightarrow \mathrm{Spec}(\mathbb{Z}) \ .$$

We obtain an affine scheme over $\mathbb{Z}$. Now $\mathrm{Spec}(\mathbb{Z})$ is a one-dimensional base, the fibres are one-dimensional curves, and $\mathrm{Spec}(R)$ is two-dimensional. It is an arithmetic surface. We want to study the fibres in more detail.

Let $W \rightarrow Z$ be a homomorphism of arbitrary schemes and p a point on the base scheme Z. The topological fibre over p is just the usual pre-image of the point p. But here we have to give the fibre the structure of a scheme. The general construction is as follows. Represent the point p by its residue field $k(p)$ and a homomorphism of schemes $\mathrm{Spec}(k(p)) \rightarrow Z$. Take the "fibre product of schemes" of the scheme W with $\mathrm{Spec}(k(p))$ over Z.

Instead of giving the general definition let me just write this down in our affine situation:

$$
\begin{array}{ccc}
\mathrm{Spec}(R) & \longleftarrow & \mathrm{Spec}(R \underset{\mathbb{Z}}{\otimes} k(p)) \\
\downarrow & & \downarrow \\
\mathrm{Spec}(\mathbb{Z}) & \longleftarrow & \mathrm{Spec}(k(p))
\end{array}
\qquad
\begin{array}{ccc}
R & \longrightarrow & R \underset{\mathbb{Z}}{\otimes} k(p) \\
\uparrow & & \uparrow \\
\mathbb{Z} & \longrightarrow & k(p)
\end{array}
$$

Both diagrams are commutative diagrams and are dual to each other.

Here we obtain for the generic point $[0]$ the residue field $k(0) = \mathbb{Q}$ and as fibre the Spec of

$$R \underset{\mathbb{Z}}{\otimes} \mathbb{Q} = \mathbb{Q}[X,Y]/(X^2 - Y^2 - 5) \ .$$

For the closed points $[p]$ we get $k(p) = \mathbb{F}_p$ and as fibre the Spec of

$$R \underset{\mathbb{Z}}{\otimes} \mathbb{F}_p = \mathbb{F}_p[X,Y]/(X^2 - Y^2 - 5) \ .$$

In the fibres over the primes we just do calculation modulo p. A point lying on a curve in the plane is a singular point of the curve if both partial derivatives of the defining equation vanish at this point. Zero conditions for functions are always closed conditions. Hence nonsingularity is an open condition on the individual curve. In fact, it is even an open condition with respect to the variation of the point on the base scheme. The curve $X^2 - Y^2 - 5 = 0$ is a nonsingular curve over $\mathbb{Q}$. The openness principle applied to the base scheme says that there are only finitely many primes for which the fibre will become singular. Here it is quite easy to calculate these primes. Let $f(X,Y) = X^2 - Y^2 - 5$ be the defining equation. Then $\partial f/\partial X = 2X$ and $\partial f/\partial Y = 2Y$. For $p = 2$ both partial derivatives vanish at every point on the curve (the fibre). Hence every point of the fibre is a singular point. This says that the fibre over the point $[(2)]$ is a multiple fibre. In this case we see immediately $(X^2 - Y^2 - 5) \equiv (X+Y+1)^2 \mod 2$. This special fibre is $\mathrm{Spec}(\mathbb{F}_2[X,Y]/((X + Y + 1)^2)$ which is a nonreduced scheme. For $p \neq 2$ the only candidate for a singular point is $(0,0)$. But this candidate lies on the curve if and only if

$5 \equiv 0 \mod p$, hence only for $p = 5$. In this case we get one singularity. Here we calculate that $(X^2 - Y^2 - 5) = (X + Y)(X - Y) \mod 5$. Altogether we obtain that nearly every fibre is a nonsingular conic. Only the fibre over $[(2)]$ is a double line and the fibre over $[(5)]$ is a union of two lines which meet at one point.

12.4.4 Other Objects

Earlier we already saw that moduli problems (degenerations, etc.) can be conveniently described as functors. One of the reasons for introducing schemes was that in many cases there does not exist varieties representing moduli functors. If we allow schemes to represent them the situation is much better. But certain moduli functors are not even representable by schemes. To obtain a representing geometric object it is sometimes necessary to enlarge the category of schemes by introducing more general objects like algebraic spaces and algebraic stacks. In a first step it is necessary to introduce a finer topology on the schemes, the *etale* topology. With respect to the etale topology one has more open sets. Schemes are "glued" together from affine schemes using algebraic morphisms. Algebraic spaces are objects where the "glueing maps" are more general maps (etale maps). Algebraic stacks are even more general than algebraic spaces. The typical situation where they occur is in connection with moduli functors. Here one has a scheme which represents a set of certain objects. If one wants to have only one copy for each isomorphy class of the objects one usually has to divide out a group action. But not every orbit space of a scheme by a group action can be made to a scheme again. Hence we indeed get new objects. These new objects are the algebraic stacks.

You can find more information on algebraic spaces in the book of Artin [Ar] or Knutson [Kn]. For stacks the appendix of [Vi] gives a very short introduction and some examples.

Hints for Further Reading

[Ar]   Artin, M., *Algebraic Spaces*, Yale Mathematical Monographs **3**, Yale University Press, 1971.

[EGA] Grothendieck, A., Dieudonné, J.A., *Eléments de géométrie algébrique I*, Springer, 1971.

[EGAA] Grothendieck, A., Dieudonné, J.A., *Eléments de géométrie algébrique*, Publications Mathématiques de l'Institute des Hautes Études Scientifiques, **8, 11, 17, 20, 24, 28, 32**.

[EHS] Eisenbud, D., Harris, J., *Schemes: The Language of Modern Algebraic Geometry*, Wadsworth & Brooks, Pacific Grove, California, 1992.

[Ha]   Hartshorne, R., *Algebraic Geometry*, Springer, 1977.

[Kn]   Knutson, D., *Algebraic Spaces*, Lecture Notes in Mathematics **203**, Springer, 1971.

[Ku]   Kunz, E., *Einführung in die kommutative Algebra und algebraische Geometrie*, Vieweg, Braunschweig, 1979; (in English translation) *Introduction to Commutative Algebra and Algebraic Geometry*, Birkhäuser, Boston, 1985.

[Mred] Mumford, D., *The Red Book of Varieties and Schemes*, Lecture Notes in Mathematics **1358** (reprint), Springer, 1988.

[M]   Mumford, D., Stability of Projective Varieties, *L'Enseign. Math.* **23** (1977) 39–110.

[Mpg]   Mumford, D., Picard Groups of Moduli Problems, (in) *Arithmetical Algebraic geometry* (Purdue 1963), O.F.G. Schilling (ed.), 33–38, Harper & Row, New York, 1965.

[Vi]   Vistoli, A., Intersection Theory on Algebraic Stacks and Their Moduli Spaces, Appendix, *Invent. Math.* **97** (1989) 613–670.

13

Hodge Decomposition and Kähler Manifold

13.1 Some Introductory Remarks on Mirror Symmetry

The goal of the following two sections is to give a short glance on the mathematics involved in the famous mirror symmetry, or more precisely mirror conjecture, originating from string theory.

In very rough terms, its origin in physics might be described as follows. There are five models for superstring theories. Their common feature is that the dimension of spacetime equals ten. To obtain the four-dimensional spacetime, which we experience, the remaining six dimensions are thought to be very small and compact. This splitting is called *string compactification*. The geometry of the six-dimensional compact manifold will be responsible for the observed physics in four dimensions. As a first consequence the observed physics in four dimensions restricts the kind of manifolds which are possible as compact manifolds. The manifolds should be special Kähler manifolds (they will be introduced in Chap. 13). As candidates are discussed:

(1) complex tori,
(2) complex tori divided by finite group actions. These quotients might have singularities and they are not necessarily manifolds again. They are called orbifolds.
(3) Calabi-Yau 3-manifolds in the strong sense, as they will be introduced in Sect. 14.1.

Depending on the situation, one speaks about toroidal, orbifold or Calabi-Yau (CY) compactification. From supersymmetry considerations, for physicists the CY compactification is the most appealing one. I will introduce them in the next section. For the moment it is enough to know that a CY threefold is a compact complex three-dimensional manifold with trivial canonical bundle. This implies that there exists a nowhere vanishing holomorphic top differential form.

Also mixed forms of compactifications are possible. A special role in the mixed cases is played by K3 (complex) surfaces. They are Calabi-Yau manifolds of dimension 2.

M. Schlichenmaier, Hodge Decomposition and Kähler Manifold. In: M. Schlichenmaier,
An Introduction to Riemann Surfaces, Algebraic Curves and Moduli Spaces, Theoretical
and Mathematical Physics, 169–182 (2007)
DOI 10.1007/978-3-540-71175-9_14

As a second consequence, the physical models obtained by different compactification show certain dualities. These dualities of the physical model should correspond to certain dualities between the compactified spaces. In the Calabi-Yau compactification case such a duality is the mirror symmetry. For complex dimension three it relates type IIA string theory[1] compactified on one CY manifold X with a type IIB string theory compactified on a CY manifold Y called its mirror. For K3 surfaces the type IIA of X and of its mirror Y and their corresponding type IIB are related. The physical mirror symmetry implies that there should be a relation between certain geometric objects of X and certain geometric objects of its mirror Y.

In string theory the fundamental objects are the maps from world-sheets, given by Riemann surfaces Σ, to spacetime. If we consider only the compact factor we obtain $\sigma : \Sigma \to X$ with X, e.g. a Calabi-Yau threefold. The appearing theories are called σ-models. The type IIA and type IIB string theories correspond to two different σ-models, the A-model and the B-model. Mirror symmetry gives a duality between these models.

It is a mathematical challenge not only to prove such a relation but also to give a mathematical explanation of it. Indeed quite a number of mathematically unexpected results have been conjectured. Ground-breaking was that physicists gave a relation between the generator function for the numbers of genus zero curves of degree d on one Calabi-Yau threefold X and the period integrals of the top-dimensional holomorphic form on its mirror, which allows to calculate these numbers for small degree, see Sect. 14.4.2.

From the mathematical point of view there are different levels of mirror symmetry, respectively mirror conjectures. Some of these relations are proven for certain types of Calabi-Yau manifolds, some are only speculative. What is definitely true is that it is a field of ongoing active mathematical research.

The first variant of mirror symmetry is the conjecture that for certain classes of CY manifolds X there exist mirrors Y such that the Hodge numbers (see (13.31)) fulfil the relation

$$h^{p,q}(X) = h^{n-p,q}(Y) . \tag{13.1}$$

Sometimes this is called *geometric mirror symmetry*.

A second variant is the *numerical mirror symmetry* which gives a relation between the generating function for the number of rational curves of degree d on X and a function defined in terms of period integrals of the mirror of X. The two objects are correlators of the A- and the B-models, respectively. And mirror symmetry says that the correlators are related in a certain manner.

The third variant is *Kontsevich's homological mirror conjecture*. It says that for mirror pairs (X, Y) of Calabi-Yau manifolds the derived Fukaya A_∞-category of Lagrangian submanifolds of X is isomorphic to the derived category of coherent sheaves of Y.

---

[1] The reader who does not know what "type IIA string theory", etc. are should just take them as names. They will not be needed in the following.

A further aspect of mirror symmetry is the Strominger–Yau–Zaslov conjecture. It is based on open string theory and D-branes. The expected duality can be expressed in terms of compatible fibrations of special Lagrangian tori.

In any case it might be too much to expect that for every CY manifold X there exists a mirror. A part of a positive solution of the mirror conjecture is to determine the class of CY manifolds admitting mirrors. The geometric mirror conjecture was proved by Batyrev for Calabi-Yau complete local intersections in toric varieties, see Sect. 14.3.2 for more details. The numerical mirror conjecture was proved by Givental, again for the same class of Calabi-Yau manifolds, see Sect. 14.4.2 for more details.

Since mirror symmetry was formulated, quite a number of books and surveys of the different variants of mirror symmetry exist. A partial list can be found at the end of Chap. 14. It is neither my goal nor my competence to give here another survey. My only aim is to introduce the mathematical concepts necessary to understand at least some of the formulations and ideas in mirror symmetry. In detail we will study the Hodge numbers and discuss the aspect that for a mirror pair (X, Y) the moduli of the deformation of the complex structure of X is in correspondence to the moduli of the deformation of the Kähler structure on the mirror Y. This corresponds to the identity $h^{2,1}(X) = h^{1,1}(Y)$ of the Hodge numbers. In this section Hodge numbers and Kähler manifolds are studied. In Chap. 14 Calabi-Yau manifolds are introduced and the mirror conjectures are defined in precise terms.

13.2 Compact Complex Manifolds and Hodge Decomposition

In Sect. 1.2 we introduced complex manifolds. Recall that a complex manifold M of dimension n has local complex coordinates $z_1, z_2, \ldots, z_n$. Two sets of local coordinates are related by holomorphic transition functions with nonvanishing holomorphic functional determinant. In detail: let $w_1, w_2, \ldots, w_n$ be a second set of holomorphic coordinates, then on the common intersection where both set of coordinates are defined we get that

$$w_j = w_j(z_1, z_2, \ldots, z_n), \qquad j = 1, 2, \ldots, n \qquad (13.2)$$

are holomorphic functions, and that

$$\frac{\partial(w_1, w_2, \ldots, w_n)}{\partial(z_1, z_2, \ldots, z_n)} \neq 0. \qquad (13.3)$$

Fundamental examples of complex manifolds are

(1) the complex number space $\mathbb{C}^n$ itself,
(2) the complex projective space $\mathbb{P}^n = (\mathbb{C}^{n+1} \setminus \{0\})/\sim$, see Example 1.4 (the coordinates are the affine coordinates),

(3) Riemann surfaces,
(4) the n-dimensional complex tori introduced in Chap. 5,
(5) smooth projective or affine varieties over $\mathbb{C}$.

In Chap. 4 tangent vectors and differentials were introduced. For a real manifold with coordinates $u_1, u_2, \ldots, u_m$ a 1-differential μ can be locally given by

$$\mu = \sum_{i=1}^{m} f_i(u) du_i , \qquad u = (u_1, u_2, \ldots, u_m) , \qquad (13.4)$$

with local functions f_i. These local functions transform under the change of coordinates in such a way that they compensate for the change of the du_i such that μ remains unchanged.

Higher differentials are given as higher exterior powers. In other words a k-differential (also simply called k-form) is locally given as

$$\sum_{i_1 < i_2 < \cdots < i_k} f_{i_1 i_2 \cdots i_k}(u) du_{i_1} \wedge du_{i_2} \wedge \cdots du_{i_k} , \qquad (13.5)$$

again with a corresponding transformation law for the local functions $f_{i_1 i_2 \cdots i_k}$.

For our complex manifolds with complex dimension n and real dimension $m = 2n$ we get for $j = 1, 2, \ldots, n$

$$z_j = x_j + i y_j , \quad \bar{z}_j = x_j - i y_j , \quad dz_j = dx_j + i dy_j , \quad d\bar{z}_j = dx_j - i dy_j . \quad (13.6)$$

This implies

$$dx_j = \frac{1}{2}(dz_j + d\bar{z}_j) , \qquad dy_j = \frac{1}{2i}(dz_j - d\bar{z}_j) , \qquad j = 1, 2, \ldots, n . \quad (13.7)$$

Hence the differentials of all orders can be expressed in terms of

$$dz_j \quad \text{and} \quad d\bar{z}_j , \quad j = 1, 2, \ldots, n . \qquad (13.8)$$

In particular, a general complex 1-differential can be written as

$$\mu = \sum_{i=1}^{n} f_i(z) dz_i + \sum_{i=1}^{n} g_i(z) d\bar{z}_i , \qquad (13.9)$$

with local complex-valued functions f_i and g_i. Be aware that the f_i and g_i are not required to be holomorphic.

Definition 13.1. *Given a coordinate patch with local holomorphic coordinates $z_1, z_2, \ldots, z_n$ a k-differential form ψ is called of type $(r, k - r)$ with $0 \leq r \leq k$ if ψ can be written locally as*

$$\psi = \sum_{\substack{i_1 < i_2 < \ldots < i_r \\ j_1 < j_2 < \ldots < j_{k-r}}} f_{i_1 i_2 \ldots i_r, j_1 j_2 \ldots j_{k-r}}(z) dz_{i_1} \wedge dz_{i_2} \wedge \cdots dz_{i_r} \wedge d\bar{z}_{j_1} \wedge \cdots d\bar{z}_{j_{k-r}} .$$

$$(13.10)$$

As the transition functions between two sets of holomorphic coordinates are holomorphic the following proposition can easily be verified.

Proposition 13.2. *The type of a differential form is well defined, i.e. it does not depend on the holomorphic coordinates chosen.*

As in the case of 1-differentials the k-differentials define sheaves. In fact they are locally free sheaf of modules (of finite rank) over the sheaf of differentiable functions.[2] Let $\mathcal{E}^k$ be the sheaf of (complex-valued) k-forms. Denote by $\mathcal{E}^{r,s}$ the sheaf of forms of type (r, s). Clearly

$$\mathcal{E}^k = \bigoplus_{r=0}^{k} \mathcal{E}^{r,k-r} . \tag{13.11}$$

Moreover from the very definition

$$\mathcal{E}^k = 0 , \quad \text{for } k > 2n , \qquad \mathcal{E}^{r,s} = 0 , \quad \text{if } r > n \text{ or } s > n . \tag{13.12}$$

If we consider M as a real differentiable manifold we denote the real tangent bundle by TM and by T^*M the real cotangent bundle. It is a differentiable bundle of rank $m = 2n$ over $\mathbb{R}$. Let $T_{\mathbb{C}}^*M$ be the complexified cotangent bundle, i.e. the bundle with local sections given by the complex-valued differentials. It is a differentiable bundle of rank $4n$ over $\mathbb{R}$ and rank $2n$ over $\mathbb{C}$. Furthermore let T_h^*M be the holomorphic cotangent bundle. Its sections are the complex-valued differential forms of type $(1, 0)$. Its real rank is $2n$, and its complex rank n.

As usual let $\bigwedge^k E$ be the kth exterior power of a vector bundle E. Then $\mathcal{E}^k$ is the sheaf of differentiable sections associated to $\bigwedge^k T_{\mathbb{C}}^*M$ and $\mathcal{E}^{k,0}$ is the sheaf of differentiable sections associated to $\bigwedge^k T_h^*M$.

A warning is here in order. The bundle $T_{\mathbb{C}}^*M$ and all its exterior powers are only differentiable bundles, not holomorphic bundles. Hence it does not make sense to talk about holomorphic sections. On the contrary T_h^*M and its exterior powers are also holomorphic bundles, hence we can talk about holomorphic sections. Their sheaves of local sections are denoted by Ω^k. They are locally free sheaves of finite rank over the sheaf $\mathcal{O}$ of holomorphic functions. Obviously Ω^k is a subsheaf of $\mathcal{E}^{k,0}$.

Next we describe the de Rham and the Dolbeault differential. For Riemann surfaces they have been already discussed in Sect. 4.2. The definition there can be extended to arbitrary dimension. First given $f \in \mathcal{E}^0(U)$, i.e. a local function f then we define

$$\partial f = \sum_{i=1}^{n} \frac{\partial f}{\partial z_i}(z)dz_i , \qquad \bar{\partial} f = \sum_{i=1}^{n} \frac{\partial f}{\partial \bar{z}_i}(z)d\bar{z}_i , \qquad df = (\partial + \bar{\partial})f . \tag{13.13}$$

---

[2] We recall our general convention that *differentiable* means *infinitely many often differentiable.*

Let $\psi \in \mathcal{E}^k(U)$ be given as

$$\psi = f(z)\psi_0, \qquad \psi_0 = dz_{i_1} \wedge \cdots \wedge dz_{i_r} \wedge d\bar{z}_{j_1} \wedge \cdots \wedge d\bar{z}_{i_{k-r}}$$

then we define

$$\partial\psi := (\partial f) \wedge \psi_0 , \qquad \bar{\partial}\psi := (\bar{\partial}f) \wedge \psi_0 , \qquad d\psi = (\partial + \bar{\partial})\psi . \qquad (13.14)$$

For a general element of $\mathcal{E}^k(U)$ the operations are defined by linear extension.

Proposition 13.3. *The above defined operators are operators on the following spaces:*

$$d : \mathcal{E}^k \to \mathcal{E}^{k+1} , \quad \partial : \mathcal{E}^{r,s} \to \mathcal{E}^{r+1,s} , \quad \bar{\partial} : \mathcal{E}^{r,s} \to \mathcal{E}^{r,s+1} , \qquad (13.15)$$

and they fulfil

$$d \circ d = \partial \circ \partial = \bar{\partial} \circ \bar{\partial} = 0$$
$$\partial \circ \bar{\partial} + \bar{\partial} \circ \partial = 0 . \qquad (13.16)$$

The proof is a direct calculation. □

The operator d is called the *de Rham* operator or *exterior differential*. The operator $\bar{\partial}$ is called the *Dolbeault operator*, or *deBar* operator.

Based on this proposition we have the following complexes of sheaves

$$0 \longrightarrow \mathbb{C} \longrightarrow \mathcal{E}^0 \xrightarrow{d} \mathcal{E}^1 \xrightarrow{d} \mathcal{E}^2 \cdots \xrightarrow{d} \mathcal{E}^{2n} \longrightarrow 0 , \qquad (13.17)$$

$$0 \longrightarrow \Omega^p \longrightarrow \mathcal{E}^{(p,0)} \xrightarrow{\bar{\partial}} \mathcal{E}^{(p,1)} \xrightarrow{\bar{\partial}} \mathcal{E}^{(p,2)} \cdots \xrightarrow{\bar{\partial}} \mathcal{E}^{(p,n)} \longrightarrow 0 , \qquad (13.18)$$

The first maps are always the inclusions, i.e. $\mathbb{C}$ is embedded as constant function, etc.

Recall that a sequence of abelian groups F_i (or vector spaces, or sheaves, ...) with morphisms ψ_i

$$\longrightarrow F_{-1} \longrightarrow F_0 \xrightarrow{\psi_0} F_1 \xrightarrow{\psi_1} \cdots \xrightarrow{\psi_{k-1}} F_k \xrightarrow{\psi_k} \cdots \qquad (13.19)$$

is called a complex if

$$\psi_k \circ \psi_{k-1} = 0 , \qquad \forall k \in \mathbb{Z} . \qquad (13.20)$$

Equivalent is the fact that

$$\operatorname{im} \psi_{k-1} \subseteq \ker \psi_k , \qquad \forall k \in \mathbb{Z} . \qquad (13.21)$$

A complex is called exact if we have equality, i.e.

$$\operatorname{im} \psi_{k-1} = \ker \psi_k , \qquad \forall k \in \mathbb{Z} . \qquad (13.22)$$

Given such a complex its cohomology objects[3] are defined as

$$\mathcal{H}^k = \frac{\ker \psi_k}{\operatorname{im} \psi_{k-1}} . \tag{13.23}$$

Consequently, if the complex is exact all its homology objects will vanish.

By Proposition 13.3 the sequences of sheaves (13.17) and (13.18) are indeed complexes. Moreover they are exact because locally the corresponding closed differential forms can be "integrated". If we take the global sections of the sheaves we obtain the sequences

$$0\longrightarrow\mathbb{C}\longrightarrow\mathcal{E}^0(M)\stackrel{d}{\longrightarrow}\mathcal{E}^1(M)\stackrel{d}{\longrightarrow}\mathcal{E}^2(M)\cdots\stackrel{d}{\longrightarrow}\mathcal{E}^{2n}(M)\longrightarrow0 , \tag{13.24}$$

$$0\longrightarrow\Omega^p(M)\longrightarrow\mathcal{E}^{(p,0)}(M)\stackrel{\bar{\partial}}{\longrightarrow}\mathcal{E}^{(p,1)}(M)\cdots\stackrel{\bar{\partial}}{\longrightarrow}\mathcal{E}^{(p,n)}(M)\longrightarrow0 . \tag{13.25}$$

Clearly these sequences remain complexes but in general they will not be exact anymore. Denote the cohomology groups of these complexes by

$$\mathrm{H}^k_{dR}(M) = \frac{\ker(d : \mathcal{E}^k(M) \to \mathcal{E}^{k+1}(M))}{\operatorname{im}(d : \mathcal{E}^{k-1}(M) \to \mathcal{E}^k(M))} , \tag{13.26}$$

and

$$\mathrm{H}^{p,k}_{\bar{\partial}}(M) = \frac{\ker(\bar{\partial} : \mathcal{E}^{(p,k)}(M) \to \mathcal{E}^{(p,k+1)}(M))}{\operatorname{im}(\bar{\partial} : \mathcal{E}^{(p,k-1)}(M) \to \mathcal{E}^{(p,k)}(M))} . \tag{13.27}$$

The $\mathrm{H}^k_{dR}(M)$ are called the de Rham cohomology groups and $\mathrm{H}^{p,k}_{\bar{\partial}}(M)$ the Dolbeault cohomology groups.

Sheaf cohomology was defined in Sect. 8.3. We are interested in the sheaf cohomology of $\mathbb{C}$ and Ω^p. The sequences (13.17) and (13.18) are *resolutions* for them. Moreover the sheaves $\mathcal{E}^k$ and $\mathcal{E}^{(r,s)}$ are *fine sheaves* and by general arguments in cohomology theory the cohomology groups of the sheaves $\mathbb{C}$ and Ω^p can be calculated via these resolutions, by taking the cohomology of the complexes (13.24) and (13.25). This yields

Proposition 13.4.

$$\mathrm{H}^k(M,\mathbb{C}) \cong \mathrm{H}^k_{dR}(M) , \tag{13.28}$$

$$\mathrm{H}^k(M,\Omega^p) \cong \mathrm{H}^{p,k}_{\bar{\partial}}(M) . \tag{13.29}$$

Note that $\mathrm{H}^k(M,\mathbb{C})$ coincides with the usual cohomology of M as topological space. It can be calculated via simplicial homology (see Sect. 2.2). By considering $\mathrm{H}^k(M,\mathbb{C})$ instead of $\mathrm{H}^k(M,\mathbb{Z})$ one takes complex-valued cocycles and hence ignores torsion elements.

---

[3] Here we consider complexes with increasing indices. Hence we obtain cohomology objects. For complexes with decreasing indices one obtains homology objects. Sometimes complexes with increasing indices are called cocomplexes.

Definition 13.5. *Assume that the cohomology spaces are finite-dimensional. The Betti numbers $b_k(M)$ are defined as*

$$b_k(M) := \dim \mathrm{H}^k(M, \mathbb{C}) = \dim \mathrm{H}^k_{dR}(M) . \qquad (13.30)$$

The Hodge numbers $h^{p,q}(M)$ are defined as

$$h^{p,q}(M) := \dim \mathrm{H}^q(M, \Omega^p) = \dim \mathrm{H}^{p,q}_{\bar\partial}(M) . \qquad (13.31)$$

The Euler characteristic (sometimes also called Euler number) is defined as

$$\chi(M) = \sum_{k=0}^{\dim_{\mathbb{R}} M} (-1)^k b_k . \qquad (13.32)$$

Theorem 13.6. *Let M be a compact complex manifold then the cohomology spaces $\mathrm{H}^k(M, \mathbb{C})$ and $\mathrm{H}^q(M, \Omega^p)$ are finite dimensional.*

As $\mathrm{H}^q(M, \Omega^p)$ is concerned this is a deep theorem. One uses elliptic complexes to show that there is a unique harmonic representative in every cohomology class. The corresponding theory is called *Hodge theory*, see [We] and [GH].

The Hodge and Betti numbers are exactly the numbers which play a role in mirror symmetry. But first we consider duality relations valid for every compact complex manifolds.

Theorem 13.7. *(Poincaré duality) Let M be a compact orientable manifold of real dimension m, then*

$$\mathrm{H}^{m-r}(M, \mathbb{C}) \cong \mathrm{H}_r(M, \mathbb{C}) \cong \mathrm{H}^r(M, \mathbb{C}) , \qquad r = 0, \dots, m . \qquad (13.33)$$

The proof can be found in every book on algebraic topology. See also [We, p. 169] for an analytic proof.

The other duality we have is Serre duality already encountered several times in this book, see, e.g. (8.11). We need a slight generalization (also to be found in [We, p. 170]).

Theorem 13.8. *(Serre duality) Let M be a compact complex manifold of dimension n and E a locally free sheaf (resp. a vector bundle) then*

$$\mathrm{H}^k(M, \Omega^p \otimes E)^* \cong \mathrm{H}^{n-k}(M, \Omega^{n-p} \otimes E^*) , \quad k, p = 0, 1, \dots, n . \qquad (13.34)$$

If we take for E the trivial bundle then from both theorems the following follows:

Proposition 13.9. *Let M be a compact complex manifold of dimension n. Then*

$$b_r(M) = b_{2n-r}(M) , \qquad r = 0, 1, \dots, 2n , \qquad (13.35)$$

$$h^{p,q}(M) = h^{n-p, n-q}(M) . \qquad (13.36)$$

In the next section we will introduce Kähler manifolds. For Kähler manifolds there exist further relations between these numbers.

13.3 Kähler Manifolds

Let M be a complex manifold. A 2-form ω of M is called a *Kähler form* (compatible with the complex structure of M) if

(1) ω is a $(1,1)$-form, i.e. it can be written locally as

$$\omega = \mathrm{i} \sum_{i,j=1}^{n} g_{ij}(z) dz_i \wedge d\bar{z}_j , \tag{13.37}$$

(2) ω is closed, i.e. $d\omega = 0$,
(3) ω is positive definite, i.e. the matrix $(g_{ij}(z))$ is a positive definite hermitian matrix for every z. In particular, the form ω is non degenerate.

Definition 13.10. *A Kähler manifold is a pair (M, ω) with a complex manifold M and ω a Kähler form on M. A complex manifold is called Kähler type if M admits at least one Kähler form.*

In the following we will give examples of Kähler manifolds, resp. of manifolds of Kähler type:

Example 13.11. The complex number space $\mathbb{C}^n$ endowed with the form

$$\omega = \mathrm{i} \sum_{j=1}^{n} dz_j \wedge d\bar{z}_j , \tag{13.38}$$

is a Kähler manifold.

Example 13.12. The complex projective line $\mathbb{P}^1$ (e.g. the compact Riemann surface of genus zero) with Kähler form

$$\omega = \frac{\mathrm{i}}{(1 + z\bar{z})^2} dz \wedge d\bar{z} . \tag{13.39}$$

Here z is the quasi-global coordinate.

Example 13.13. The complex projective space $\mathbb{P}^n$ with the Fubini–Study form

$$\omega_{FS} = \mathrm{i} \frac{(1 + |z|^2) \sum_{i=1}^{n} dz_i \wedge d\bar{z}_i - \sum_{i,j=1}^{n} \bar{z}_i z_j dz_i \wedge d\bar{z}_j}{(1 + |z|^2)^2} . \tag{13.40}$$

Here the z_i are affine coordinates in an affine chart.

Example 13.14. Every Riemann surface is of Kähler type. On Riemann surfaces we always have volume forms. Each volume form can be expressed as $f(x, y)dx \wedge dy$ (x and y the real coordinates). But using (13.7) we see that the volume form can be rewritten as $g(z)dz \wedge d\bar{z}$. As top form it is already closed and as volume form it is nondegenerate and positive.

Example 13.15. Recall from Sect. 5.1 that a n-dimensional complex torus T is given as a quotient of $\mathbb{C}^n$ by a lattice L in $\mathbb{C}^n$, i.e. $T = \mathbb{C}^n/L$. As the lattice is discrete the quotient is a complex manifold and local complex coordinates are given by the local coordinates z_i for $i = 1, \ldots, n$ of $\mathbb{C}^n$. Obviously the form (13.38) is invariant under the lattice translation, hence it descends to the quotient. As the conditions for a form to be a Kähler form are only local conditions, it will define a Kähler form on the torus. Hence every n-dimensional complex torus will be a Kähler manifold. If $n \geq 2$ then not all tori will be projective manifolds, as not all of them admit enough theta functions to embed the torus into projective space, see Sect. 6.3.

Example 13.16. In Chap. 7 the upper half-plane

$$\mathcal{H} := \{\, z \in \mathbb{C} \mid \operatorname{Im} z > 0 \,\}$$

has been introduced. Clearly it is a complex manifold and as a open subset of $\mathbb{C}$ it carries the Kähler structure coming from $\mathbb{C}$. But in fact there is another Kähler form more suitable for it:

$$\omega := \frac{i}{(\mathbf{Im}\, z)^2} dz \wedge d\bar{z} \,. \tag{13.41}$$

Why is it more adapted? On $\mathcal{H}$ the group $\mathrm{PSL}(2,\mathbb{R})$ is operating via fractional linear transformations (7.5). It is easy to verify that (13.41) is invariant under the group action. This has an important consequence. As explained in Sect. 7.2 every compact Riemann surface X of genus $g \geq 2$ can be obtained as quotient of $\mathcal{H}$ by a certain subgroup of $\mathrm{PSL}(2,\mathbb{R})$ (a Fuchsian group). By the invariance the Kähler form descends to X. This endows X with a special Kähler structure. If one takes the metric associated to this Kähler structure it has constant scalar curvature -1.

Starting from the above examples we get many more by

Proposition 13.17. *Complex submanifolds of Kähler manifolds are Kähler manifolds by restricting the Kähler form.*

For the proof see [We, p. 190].

As the projective space is a Kähler manifold and projective manifolds are submanifolds of $\mathbb{P}^n(\mathbb{C})$ we obtain

Corollary 13.18. *Projective manifolds are of Kähler type.*

Remark 13.19. All complex manifolds which appeared in the book up to now are Kähler manifolds. But it should be kept in mind that not every complex manifold is a Kähler manifold.

Remark 13.20. If we multiply the Kähler form by a positive real constant we will obtain another Kähler form. We do not consider these two Kähler forms to be essentially different. Nevertheless it might be the case that the same complex manifold has more than one Kähler form which are essentially different. But given two Kähler forms ω_1 and ω_2 their sum $\omega_1 + \omega_2$ is again a Kähler form. This is obvious from the definition. The set of all Kähler forms on M is a real cone in the space of all forms.

For compact Kähler manifolds the cohomology spaces have additional internal algebraic structures.

Theorem 13.21. *Let M be a compact complex manifold of Kähler type. Then there exists a direct sum decomposition*

$$\mathrm{H}^k(M, \mathbb{C}) = \sum_{p+q=k} \mathrm{H}_{\bar{\partial}}^{p,q}(M) = \sum_{p+q=k} \mathrm{H}^q(M, \Omega^p) , \qquad (13.42)$$

and

$$\overline{\mathrm{H}}_{\bar{\partial}}^{p,q}(M) = \mathrm{H}_{\bar{\partial}}^{q,p}(M) . \qquad (13.43)$$

Based on the decomposition, sometimes we write $\mathrm{H}^{q,p}(M, \mathbb{C})$ for $\mathrm{H}_{\bar{\partial}}^{q,p}(M)$.

As a consequence of the theory we can collect the following relations between the Betti numbers and the Hodge numbers (using also Proposition 13.9).

Proposition 13.22.

$$b_k(M) = \sum_{p+q=k} h^{p,q}(M) , \qquad (13.44)$$

$$h^{p,q}(M) = h^{q,p}(M) , \qquad (13.45)$$

$$h^{p,q}(M) = h^{n-p,n-q}(M) , \qquad (13.46)$$

$$b_k(M) \quad \text{is even for } k \text{ odd} , \qquad (13.47)$$

$$h^{1,0}(M) = \frac{1}{2} b_1(M) \quad \text{and hence a topological invariant.} \qquad (13.48)$$

Proof. The first two statements follow immediately from the above theorem. For k odd all Hodge numbers in the sum will appear in pairs. This implies the third statement. Finally, Statement 4 is a special case for $k = 1$.   □

Recall that $h^{1,0}(M) = \dim \mathrm{H}^0(M, \Omega)$ is the dimension of the space of holomorphic 1-differentials. The result is in perfect accordance with the case of compact Riemann surfaces of genus g. There we have exactly g linearly independent global holomorphic differentials and $2g$ linearly independent 1-cycles.

More can be said about the Hodge numbers.

Proposition 13.23. *For a compact Kähler manifold M the following is valid:*
(a) The Kähler form ω is also $\bar{\partial}$ and ∂ closed, i.e. $\bar{\partial}\omega = \partial\omega = 0$.
(b) Furthermore, we have

$$h^{p,p}(M) \geq 1 , \qquad p = 0, 1, \ldots, n \qquad (13.49)$$

A nonvanishing element in the cohomology can be given by $\omega^{\wedge p}$ (p-times the wedge of ω with itself).
(c) The forms $\omega^{\wedge p}$ are not exact.
(d) The even Betti numbers are positive.

Proof. First note that $d\omega$ is a 3-form which vanishes. Using $d = \partial + \bar{\partial}$ and Theorem 13.21 we get $0 = d\omega = \nu_1 + \nu_2$ with $\nu_1 = \partial\omega$ a $(2,1)$-form and $\nu_1 = \bar{\partial}\omega$ a $(1,2)$-form. The sum can only vanish if both terms individually vanish. This implies (a).

In particular $\omega^{\wedge p}$ is a $\bar{\partial}$-closed form, hence defines a class in $\mathrm{H}_{\bar{\partial}}^{p,p}(M)$. Assume $\omega^{\wedge p} = d\eta$, i.e. that $\omega^{\wedge p}$ is exact. We can write

$$\omega^{\wedge n} = d\eta \wedge \omega^{\wedge n-p} = d(\eta \wedge \omega^{\wedge n-p}) , \tag{13.50}$$

as ω is closed. If we integrate we get the contradiction

$$0 \neq \int_M \omega^{\wedge n} = \int_M d(\eta \wedge \omega^{\wedge n-p}) = 0 . \tag{13.51}$$

The first inequality is due to the fact that ω is nondegenerate and $\omega^{\wedge n}$ is a volume form for M. The last equality is due to Stokes' theorem 4.5. Hence $\omega^{\wedge p}$ cannot be exact. This shows (c).

Consequently, we exhibited a nonvanishing class, which shows (b). But using the decomposition of Theorem 13.21 we see (d). $\quad\square$

In a similar way the following can be shown (see [GH, p. 110])

Proposition 13.24. *Let M be a compact Kähler manifold. Let μ be a holomorphic p-form. Then μ is d-closed, i.e. $d\mu = 0$. Furthermore, if $\mu \neq 0$ it cannot be exact.*

Based on Theorem 13.21 we can write down the Hodge numbers in a diagrammatic form. This diagram is called the *Hodge diamond*. As example the Hodge diagram of a three-dimensional manifold is given.

$$
\begin{array}{ccccccc}
b_0 & & & h^{0,0} & & & \\
b_1 & & h^{1,0} & & h^{0,1} & & \\
b_2 & & h^{2,0} & h^{1,1} & h^{0,2} & & \\
b_3 & h^{3,0} & h^{2,1} & & h^{1,2} & h^{0,3} & \\
b_4 & & h^{3,1} & h^{2,2} & h^{1,3} & & \\
b_5 & & h^{3,2} & & h^{2,3} & & \\
b_6 & & & h^{3,3} & & &
\end{array}
\tag{13.52}
$$

The sum over the Hodge number over the line k gives the corresponding Betti number b_k. Poincaré duality $h^{p,r} = h^{n-p,n-r}$ (valid for all compact complex manifolds) corresponds to a reflection at the centre of the diamond. Reflection at the vertical axes ($h^{p,r} = h^{r,p}$) is a symmetry due to the Kähler structure. If we combine these two symmetries we also obtain as symmetry a reflection at the horizontal axes.

Furthermore, as our manifold is connected, $b_0 = h^{0,0} = 1$ and by Poincare duality hence also $b_n = h^{n,n} = 1$. As shown in Proposition 13.23 the numbers on the vertical axes are always positive.

If we ignore all numbers which are determined by the other numbers we obtain the following quadrant.

$$
\begin{array}{ccc}
1 & & 1 \\[2mm]
b_1 & & h^{1,0} \\[2mm]
b_2 & h^{2,0} & h^{1,1} \geq 1 \\[2mm]
b_3 & h^{3,0} & h^{2,1}
\end{array}
\tag{13.53}
$$

13.4 Hodge Numbers of the Projective Space

For further reference we calculate the Hodge numbers of the projective space. The cohomology of the projective space $\mathbb{P}^n$ is concentrated in even dimensions. More precisely, see [GH, p. 60],

$$
\mathrm{H}^m(\mathbb{P}^n, \mathbb{Z}) = \begin{cases} \mathbb{C}, & m \text{ even}, \ 0 \leq m \leq 2n, \\ 0, & m \text{ odd} . \end{cases}
\tag{13.54}
$$

This implies for the Betti numbers $b_{2k+1} = 0$ and $b_{2k} = 1$. As a consequence $h^{p,q}(\mathbb{P}^n) = 0$ if $p+q$ is odd. If $p+q$ is even then, by the Kähler condition (13.45),

$$
b_{2k}(\mathbb{P}^n) = 1 = 2 \sum_{0 \leq r < k} h^{r,2k-r} + h^{k,k} .
\tag{13.55}
$$

But this is only possible for

$$
h^{r,2k-r} = 0 , \quad 0 \leq r < k , \qquad h^{k,k} = 1 .
\tag{13.56}
$$

This shows

Proposition 13.25.

$$
h^{p,q}(\mathbb{P}^n) = \begin{cases} 0 , & p \neq q, \\ 1, & 0 \leq p = q \leq n . \end{cases}
\tag{13.57}
$$

All cohomology classes are represented by (p,p) forms. Furthermore, $\mathbb{P}^n$ does not admit nonzero global holomorphic forms.

In particular in the Hodge diamond the values along the vertical diagonal are 1, the values outside are zero. With Proposition 13.23 it follows that the cohomology is indeed generated by $\omega^{\wedge p}$, where ω is the Fubini–Study Kähler form.

Hints for Further Reading

[GH]   Griffiths, Ph., Harris, J., *Principles of Algebraic Geometry*, Wiley, New York, 1978.

[We]   Wells, R.O., *Differential Analysis on Complex Manifolds*, Springer, 1980.

14

Calabi-Yau Manifolds and Mirror Symmetry

14.1 Calabi-Yau Manifolds

As already explained in Sect. 13.1 Calabi-Yau manifolds are the most appealing candidates for string compactifications. In the following we will give their definition. But the reader should be aware that in the literature different definitions are in use.

Recall that for a complex manifold M of complex dimension n with holomorphic cotangent bundle $T_h^* M$ the canonical bundle K_M, also denoted by Ω^n, is defined to be its highest exterior power, i.e. the holomorphic line bundle

$$K_M = \Omega^n = \bigwedge^n T_h^* M \ . \tag{14.1}$$

Definition 14.1. *A compact Kähler manifold is called a Calabi-Yau manifold (in the weak sense) if its canonical line bundle K_M is trivial, i.e.*

$$K_M \cong \mathcal{O}_M \ . \tag{14.2}$$

In other words, there exists a nowhere vanishing holomorphic n-differential form.

The Chern classes of a manifold M are defined to be the Chern classes of the holomorphic tangent bundles of M. Here, depending on the situation, the Chern classes might be considered as elements of the Chow ring or of the cohomology ring, see Sect. 10.2.

We obtain for the first Chern class of a manifold M

$$c_1(M) := c_1(T_h M) = c_1(\bigwedge^n T_h M) = -c_1(\bigwedge^n T_h^* M) = -c_1(K_M) \ . \tag{14.3}$$

Recall that for a line bundle L we get for the dual bundle L^* the equality $c_1(L^*) = -c_1(L)$.

M. Schlichenmaier, Calabi-Yau Manifolds and Mirror Symmetry. In: M. Schlichenmaier,
An Introduction to Riemann Surfaces, Algebraic Curves and Moduli Spaces, Theoretical
and Mathematical Physics, 183–202 (2007)
DOI 10.1007/978-3-540-71175-9_15

Proposition 14.2. *Let M be a compact Calabi-Yau (CY) manifold of dimension n. Then*

$$c_1(M) = 0, \tag{14.4}$$

$$h^{0,n}(M) = h^{n,0}(M) = 1, \tag{14.5}$$

$$\mathrm{H}^r(M, \bigwedge^s T_h M) \cong \mathrm{H}^r(M, \Omega^{n-s}), \quad s = 0, 1, \ldots, n. \tag{14.6}$$

Proof. As K_M is trivial by the definition, $c_1(M)$ is trivial too.

From the definition it follows that there exists a nowhere vanishing holomorphic n-form μ. By Proposition 13.24 it cannot be exact. Hence $h^{n,0} \geq 1$. Let ψ be another global holomorphic n-form. As μ has no zeros the quotient ψ/μ will be a holomorphic 0-form, hence a holomorphic function. As M is compact the quotient is a constant. This shows $\mathrm{H}^0(M, \Omega^n)$ is generated by μ. In particular, $h^{n,0}(M) = 1$. As M is Kähler (see (13.45)) we have also $h^{0,n}(M) = 1$. To show (14.6) we apply Serre duality (13.34) for $E = \bigwedge^s T_h M$, $p = 0$. Using $\Omega^n = \mathcal{O}_M$ we get

$$H^r(M, \bigwedge^s T_h M) \cong H^{n-r}(M, \bigwedge^s T_h^* M) = H^{n-r}(M, \Omega^s)$$
$$\cong \mathrm{H}^r(M, \Omega^{n-s}). \tag{14.7}$$

The last isomorphy follows again from Serre duality, now applied to $E = \mathcal{O}_M$.
□

Remark 14.3. From the triviality of the canonical bundle it follows by results of Calabi that there exists a Kähler metric whose Ricci form vanishes. Hence CY manifolds are Ricci flat.

To explain what Calabi-Yau manifolds are in the strong sense, we have to first introduce coverings. This was done already in the context of Riemann surfaces (see Sect. 2.3). But for the convenience of the reader I will repeat it for the higher dimensional analytic context here again. Given a complex manifold M another manifold $\tilde{M}$ is called a *finite unramified covering* if there is a surjective analytic map $\pi : \tilde{M} \to M$ such that every point $x \in M$ has an open connected neighbourhood U such that

$$\pi^{-1}(U) = \bigcup_{j=1,\ldots,k} V_j, \quad \text{with } V_j \cap V_i = \emptyset \ (i \neq j), \quad V_j \cong U. \tag{14.8}$$

The number k is the degree of the covering. As our manifolds are always assumed to be connected, the degree does not depend on the point $x \in M$ chosen. The covering $\tilde{M}$ is called a *universal covering* if $\tilde{M}$ is simply connected. In this case the group of automorphism of $\tilde{M}$ over M (called the covering group) (i.e. those automorphisms ϕ with $\pi \circ \phi = \pi$) can be identified with the fundamental group of M. Note that the unramified covering $\tilde{M}$ of a complex manifold M will be again a complex manifold. Moreover if M is a Kähler manifold then $\tilde{M}$ is also a Kähler manifold (by pulling back the Kähler form of M). This is clear as being a Kähler form is a local condition and $\tilde{M}$ looks locally like M. If the covering is a finite covering and M is compact $\tilde{M}$ will be again compact.

It is possible to show that every Calabi-Yau manifold M admits a finite unramified covering $\tilde{M}$ such that (see [Ma1])

M is the direct product of a complex torus M_1, of a simply connected CY manifold M_2 with $h^{2,0}(M_2) = h^{0,2}(M_2) \neq 0$ and of a simply connected CY manifold M_3 with $h^{0,2}(M_3) = 0$.

Of course some of the factors might be missing.

Definition 14.4. *A Calabi-Yau manifold M (in the weak sense) is called a Calabi-Yau manifold in the strong sense if only factors of type M_3 appear in every finite unramified covering of M.*

One might add the additional property that the Calabi-Yau manifolds are projective, i.e. embeddable in projective space.

Proposition 14.5. *For a Calabi-Yau manifold M in the strong sense we have*

$$b_1(M) = 0, \quad h^{1,0}(M) = h^{0,1}(M) = 0,$$
$$b_2(M) = h^{1,1}(M), \quad h^{2,0}(M) = h^{0,2}(M) = 0 . \tag{14.9}$$

Proof. Let $\tilde{M}$ be the finite unramified covering described above. As by definition $\tilde{M} = M_3$, it is simply connected. Hence the covering group will be the fundamental group of M. As the covering is a finite covering every element of the fundamental group will be a torsion element. The homology classes of the elements of the fundamental group generate the first simplicial homology $H_1(M, \mathbb{Z})$. But in $H_1(M, \mathbb{C})$ torsion elements cannot be seen. Hence $H_1(M, \mathbb{C}) = 0$ and $H^1(M, \mathbb{C}) = 0$, consequently $b_1(M) = 0$.
From $b_1(M) = h^{1,0}(M) + h^{0,1}(M) = 0$ it follows that both terms have to vanish.
Assume $h^{2,0}(M) = \dim H^0(M, \Omega^2) \neq 0$. Hence there exists a holomorphic global 2-form different from 0. The pullback to $\tilde{M}$ remains a nonzero holomorphic global 2-form. By Proposition 13.24 we obtain a nonzero element of $H^0(\tilde{M}, \Omega^2)$, which is in contradiction to the definition of M_3. Consequently, $h^{2,0}(M) = 0$ and by the Kähler condition $h^{0,2}(M) = 0$. Finally

$$b_2(M) = h^{2,0}(M) + h^{1,1}(M) + h^{0,2}(M) = h^{1,1}(M) . \quad \square \tag{14.10}$$

Compact complex manifolds of *dimension one* are exactly the compact Riemann surfaces studied thoroughly in this book. The only compact Riemann surfaces with trivial canonical bundle are the one-dimensional complex tori, or equivalently the elliptic curves. As for them $b_1 = 2$, they are CY manifolds in the weak sense.

In *dimension two* the well-established but more involved classification theory of complex surfaces has to be used. It turns out that only complex two-dimensional tori and the so-called K3 surfaces have trivial canonical bundles. Complex tori have nonvanishing first Betti numbers. K3 surfaces have $b_1 = 0$ but have nonvanishing $h^{2,0}$. Both types are CY manifolds in the weak sense but not in the strong sense. Indeed, it turns out that K3 surfaces are simply connected. In the above decomposition they are examples of the second factor. See Sect. 14.2 for their exact definition.

Note that both in the torus and in the K3 case there exists manifolds which are not algebraic, i.e. which cannot be embedded into projective space. Hence it will make a difference if one considers all complex manifolds or only algebraic ones.

The first Calabi-Yau manifolds in the strong sense appear in *dimension three*. In Sect. 14.4.1 I will give the standard example of quintics in $\mathbb{P}^4$.

From the information we obtained up to now we can write down the Hodge diamond for a CY three-fold in the strong sense:

$$
\begin{array}{ccccccc}
1 & & & & & & 1 \\
0 & & & 0 & & 0 & \\
b_2 & & 0 & & h^{1,1} & & 0 \\
b_3 & & 1 & h^{2,1} & & h^{1,2} & & 1 \\
b_4 & & 0 & & h^{2,2} & & 0 \\
0 & & & 0 & & 0 & \\
1 & & & & & & 1
\end{array}
\tag{14.11}
$$

with the relations
$$
h^{1,2} = h^{2,1}, \qquad h^{2,2} = h^{1,1}
$$

For the Euler characteristic of CY three-folds we calculate

$$
\chi(M) = \sum_{i=0}^{6} (-1)^i b_i(M) = 2(h^{1,1}(M) - h^{2,1}(M)) .
\tag{14.12}
$$

Remark 14.6. There is another approach to the definition of Calabi-Yau manifolds which is based on holonomy. I will not follow this approach here. Nevertheless I will give some rough explanations. Let M be a (real) manifold with tangent bundle T_M and a chosen metric. Fix a point $a \in M$ and consider the tangent space $T_a M$ at the point a. Given a path in the manifold the parallel transport of vectors (i.e. of the elements of the tangent spaces) with respect to the "metric connection", associated to the metric in the tangent bundle, is well defined. If we consider loops based at the point a we can apply to every vector the parallel transport along the loop and we will end up with another vector in $T_a M$. This vector might be different from the starting vector. The corresponding operation is the holonomy operation. As parallel transport is a linear operation and linearly independent vectors will stay linearly independent the holonomy operations generate a subgroup of the general linear group, the *holonomy group*. If one considers only contractible loops one obtains the *restricted holonomy group*. If the restricted holonomy group is a subgroup of $U(n)$, then M can be equipped with the structure of a complex Kähler manifold. If it is even a subgroup of $SU(n)$ then we have $c_1(M) = 0$ and furthermore the manifold M can be equipped with a Ricci flat Kähler metric. If the restricted holonomy is exactly $SU(n)$ then it follows that $h^{2,0}(M) = 0$, i.e. there are no holomorphic two-forms.

From the requirements of physics it turns out that the restricted holonomy groups of the compactified spaces should be contained in $SU(3)$, resp. $SU(n)$.

Hence one also finds the definition that Calabi-Yau manifolds of dimension n are compact (complex) Kähler manifolds with trivial canonical bundle and restricted holonomy group the full $SU(n)$. Calabi-Yau manifolds in the strong sense are also Calabi-Yau manifolds in this sense. Tori are flat, hence its restricted holonomy will vanish, K3 surfaces (they are always simply connected) have holonomy $Sp(1)$, in another terminology they are Hyperkähler manifolds. Hence in dimension two there are no Calabi-Yau in the holonomy sense. But as $SU(1) = \{id\}$, elliptic curves are Calabi-Yau in the holonomy sense.

14.2 K3 Surfaces, Hypersurfaces and Complete Intersections

Definition 14.7. *A compact complex surface M is called a K3 surface if its canonical bundle is trivial, i.e. $K_M \cong \mathcal{O}_M$ and the first Betti number vanishes, i.e. $b_1(M) = 0$.*

Examples of a K3 surface are given by quartic nonsingular hypersurfaces in $\mathbb{P}^3$.

As this construction will also play a role in the construction of Calabi-Yau three-folds, I will explain it in a more general setting. Let $\mathbb{P}^n$ be the n-dimensional projective space. Recall from Sect. 6.1 that projective varieties can be given as the vanishing set of finitely many homogeneous polynomials in $(n+1)$ variables (or in the language of ideals developed in Chap. 11 as the vanishing set of the elements of the homogeneous ideal generated by them). A variety is called a hypersurface if it can be given by one irreducible polynomial (or equivalently if its vanishing ideal is generated by one element). The *degree of the hypersurface* is given by the degree of this polynomial. Special cases are the hyperplanes which can be given by linear polynomials.

Hypersurfaces are complex codimension one subvarieties of $\mathbb{P}^n$ and as such they are divisors. As explained in Sect. 9.1 the divisor classes are in 1:1 correspondence to the isomorphy classes of line bundles. Let $\mathcal{O}_{\mathbb{P}^n}$ be the structure sheaf of $\mathbb{P}^n$, i.e. the trivial line bundle. Denote by $\mathcal{O}_{\mathbb{P}^n}(1)$ the locally free sheaf of rank 1, resp. the line bundle, associated to the divisor class of any hyperplane.

It turns out that all hyperplanes H are linearly equivalent divisors, hence they determine the same line bundle (up to isomorphy). A section of this line bundle has as zero set one of the hyperplanes. An arbitrary section is a linear combination of the linear homogeneous polynomials $X_0, X_1, \ldots, X_n$. We use the notation:

$$\mathcal{O}_{\mathbb{P}^n}(0) := \mathcal{O}_{\mathbb{P}^n}, \qquad \mathcal{O}_{\mathbb{P}^n}(-1) := (\mathcal{O}_{\mathbb{P}^n}(1))^*,$$

$$\mathcal{O}_{\mathbb{P}^n}(d) := (\mathcal{O}_{\mathbb{P}^n}(1))^{\otimes d}, \qquad \text{(for } d > 1), \qquad (14.13)$$

$$\mathcal{O}_{\mathbb{P}^n}(d) := ((\mathcal{O}_{\mathbb{P}^n}(1))^*)^{\otimes |d|}, \qquad \text{(for } d < -1) .$$

Based on the fact that $H^0(\mathbb{P}^n, \mathcal{O}_{\mathbb{P}^n}(1))$, is given by the linear polynomials it follows that the global sections of $\mathcal{O}_{\mathbb{P}^n}(d)$, i.e. the elements of the space $H^0(\mathbb{P}^n, \mathcal{O}_{\mathbb{P}^n}(d))$, are given by the homogeneous polynomials of degree d. In particular, a hypersurface X of degree $d \geq 1$ is linearly equivalent to $d \cdot H$:

$$X \cong d \cdot H . \qquad (14.14)$$

In fact it turns out that every line bundle L over $\mathbb{P}^n$ of degree d is isomorphic to $\mathcal{O}_{\mathbb{P}^n}(d)$.

For the canonical bundle of $\mathbb{P}^n$ one can show that (see [GH, p. 146])

$$K_{\mathbb{P}^n} = \mathcal{O}_{\mathbb{P}^n}(-(n+1)) . \qquad (14.15)$$

For $n = 1$, the case of the Riemann sphere, we obtain

$$K_{\mathbb{P}^1} = \mathcal{O}_{\mathbb{P}^1}(-2) , \qquad (14.16)$$

in accordance with the fact that every meromorphic differential for the Riemann sphere has to have degree -2, i.e. to have at least two poles.

The hypersurface in $\mathbb{P}^n$ defined by the homogeneous polynomial f in $n+1$ variables will have singularities if and only if there exists a common zero of all partial derivatives of f which is different from the point $0 \in \mathbb{C}^{n+1}$, see Definition 6.3. Obviously a hypersurface given by a generic polynomial will have no singularities; it will be smooth.

Let X be a smooth hypersurface of $\mathbb{P}^n$ of degree d and $\dim X \geq 2$ (in particular we have $n \geq 3$) and let H be a hyperplane of $\mathbb{P}^n$ in generic position with respect to X. In this case $X \cap H$ will be a smooth irreducible codimension one hypersurface of X, hence a divisor. The corresponding line bundle will be denoted by $\mathcal{O}_X(1)$. It can be equivalently described as restriction of the line bundle $\mathcal{O}_{\mathbb{P}^n}(1)$ to the submanifold X.

Proposition 14.8. *Let X be a smooth hypersurface of $\mathbb{P}^n$ of degree d. Then*

(1) $K_X \cong \mathcal{O}_X(d - (n+1))$,   (14.17)

(2) $\pi_1(X) = \{0\}$,   *if* $\dim X \geq 2$,   (14.18)

(3) $H^0(X, \Omega_X^i) = \{0\}$, $0 < i < n - 1$,   (14.19)

(4) $h^{i,0}(X) = h^{0,i}(X) = 0$, $0 < i < n - 1$.   (14.20)

Recall that $\pi_1(X)$ is the fundamental group. Hence (2) says that X is simply connected. For the proof of the proposition we need the *adjunction formula* (see [GH, p. 147]).

Theorem 14.9. *Let Y be a compact complex manifold and X a smooth analytic hypersurface then*

$$K_X = (K_Y \otimes \mathcal{O}_Y(X))_{|X} .$$ (14.21)

Here (as above) $\mathcal{O}_Y(X)$ denotes the line bundle of Y associated to the divisor X and $_{|X}$ denotes the restriction to X.

Proof. (of Proposition 14.8) By the adjunction formula (14.21)

$$K_X \cong (\mathcal{O}_{\mathbb{P}^n}(-(n+1)) \otimes \mathcal{O}_{\mathbb{P}^n}(d))_{|X}$$
$$= \mathcal{O}_{\mathbb{P}^n}(d - (n+1))_{|X} = \mathcal{O}_X(d - (n+1)) . \quad (14.22)$$

Hence (1).

As a corollary from the Lefschetz hyperplane theorem [GH, p. 159] it follows if the dimension of X is ≥ 2 that X is connected and simply connected. Hence (2).

Furthermore a corollary of the proof of the Lefschetz hyperplane theorem is that the map

$$H^p(\mathbb{P}^n, \Omega_{\mathbb{P}^n}^q) \to H^p(X, \Omega_X^q) \quad \text{is an isomorphism for } p + q \leq n - 2 . \quad (14.23)$$

Hence

$$h^{p,q}(\mathbb{P}^n) = h^{p,q}(X), \qquad p + q \leq n - 2 . \quad (14.24)$$

In particular

$$h^{p,0}(\mathbb{P}^n) = h^{p,0}(X), \qquad 0 \leq p < n - 1 . \quad (14.25)$$

But for the Hodge number of the projective space we saw in Proposition 13.25 that $h^{p,0}(\mathbb{P}^n) = 0$ for $0 < p < n - 1$. This shows (3).

The relation (4) is a consequence of the Kähler duality (13.45). □

From this proposition the following is immediate:

Proposition 14.10.
*(a) A generic degree $(n+1)$ hypersurface in $\mathbb{P}^n$ will be a Calabi-Yau manifold
(in the weak sense).
(b) A generic degree $(n+1)$ hypersurface in $\mathbb{P}^n$ for $n \geq 4$ will be a Calabi-Yau
manifold in the strong sense.
(c) The generic quartics (i.e. hypersurfaces of degree 4) in $\mathbb{P}^3$ are K3 surfaces.
(d) The generic cubics (i.e. curves of degree 3) in $\mathbb{P}^2$ are elliptic curves.*

Let us return to our K3 surfaces. From the above it is clear that quartics
in $\mathbb{P}^3$ are K3 surfaces. It can be shown that every K3 surface is of Kähler type
(but not necessarily projective). Hence K3 surfaces are Calabi-Yau surfaces
(of weak type). All K3 surfaces M are deformations of each other, hence they
are diffeomorphic. As the quartic is simply connected this implies that all of
them are simply connected.

To calculate the other Hodge numbers we use the following. For complex
surfaces we have the very useful Noether formula. It can be deduced from
the Hirzebruch–Riemann–Roch Theorem 10.6 and the Gauss–Bonnet formula
[GH] relating the degree of the highest Chern class of the holomorphic tangent
bundle with the Euler characteristic $\chi(M)$:

$$\deg(c_n(M)) := \deg(c_n(T_h M)) = \chi(M) . \tag{14.26}$$

For this formula M is a compact complex manifold of dimension n. In this
case the highest Chern class $c_n(M)$ is an element of the highest cohomology,
resp. of the nth part $\mathrm{CH}^n(M)$ of the Chow ring. Recall from Sect. 10.2 that
the degree can be obtained either by taking the degree map of $\mathrm{CH}^n(M)$ or by
integrating a representing differential form for the cohomology class $c_n(M)$
over M.

Let the dimension be two. We take in (10.7) as bundle E the trivial line
bundle $\mathcal{O}_M$. Then $ch(\mathcal{O}_M) = 1$ and with the expressions for the Todd class
(10.5) we obtain immediately

$$\chi(\mathcal{O}_M) = \frac{1}{12}(\deg(c_1(M)^2) + \chi(M)) . \tag{14.27}$$

This formula should be interpreted in the following way. Take $c_1(T_h M)$ as
divisor class, take its self-intersection (which is well defined in the Chow ring)
and finally take the degree of the result. This gives $\deg(c_1(M)^2)$. This number
is also called *self-intersection number*.

Now let M be a K3 surface. As $c_1(T_h M) = c_1(\bigwedge^2 T_h M) = -c_1(K_M)$ and
in our case $K_M = \mathcal{O}_M$ we have $c_1(T_h M) = 0$. Hence also $\deg(c_1(M)^2) = 0$.
For a K3 surface we have $h^{0,0}(M) = 1$, $h^{0,1}(M) = 0$ as M is simply connected,
and $h^{0,2}(M) = 1$ by (14.5). Hence for the holomorphic Euler characteristic

$$\chi(\mathcal{O}_M) = \sum_{i=0}^{2}(-1)^i \dim \mathrm{H}^i(M, \mathcal{O}_M) = \sum_{i=0}^{2}(-1)^i h^{0,i}(M) = 2 . \tag{14.28}$$

From (14.27) we get $\chi(M) = 24$. As

$$\chi(M) = \sum_{i=0}^{4}(-1)^i b_i(M) = 4 + h^{1,1}(M)$$

we can calculate the missing number as $h^{1,1}(M) = 20$. The Hodge diamond of K3 surfaces reads as

$$
\begin{matrix}
 & & 1 & & & & 1 & & \\
 & 0 & & & & 0 & & 0 & \\
22 & & & & 1 & & 20 & & 1 \\
 & 0 & & & & 0 & & 0 & \\
 & & 1 & & & & 1 & &
\end{matrix}
\tag{14.29}
$$

For more details on K3 manifolds see [BPV].

For completeness let me write down the Hodge diamond of the two-dimensional complex torus:

$$
\begin{matrix}
 & & 1 & & & & 1 & & \\
 & 4 & & & & 2 & & 2 & \\
6 & & & & 1 & & 4 & & 1 \\
 & 4 & & & & 2 & & 2 & \\
 & & 1 & & & & 1 & &
\end{matrix}
\tag{14.30}
$$

The hypersurface construction has to be extended to obtain more Calabi-Yau manifolds. The first extension is to consider smooth *complete intersections* in $\mathbb{P}^n$. Geometrically they are smooth transversal intersections of k hypersurfaces Y_i of degree d_i for $i = 1, \ldots, k$. The dimension of X will be $n - k$. Special cases of complete intersections are the hypersurfaces. A warning is in order. Not every smooth subvariety of dimension l will be a complete intersection, i.e. can be given as the common zero set of $n - l$ polynomials.

For a complete intersection the type is defined to be the k-tuple of degrees $(d_1, d_2, \ldots, d_k)$ of the hypersurfaces. It turns out that in generalizations of (14.17) one obtains

$$K_X = \mathcal{O}_X\left(\sum_{i=1}^{k} d_i - (n+1)\right).
\tag{14.31}$$

Hence numbers n, k, and d_i, for $i = 1, \ldots k$, can be found such that the canonical bundle will be trivial.

Furthermore, the properties in Proposition 14.8, suitable adjusted, are also valid. Complete intersections in weighted projective space and more generally in toric varieties (defined further down) give additional examples of Calabi-Yau manifolds.

14.3 Geometric Mirror Symmetry

14.3.1 The Formulation

As explained in Sect. 13.1 the geometric mirror conjecture claims that to every Calabi-Yau manifold X another Calabi-Yau manifold Y can be found such that the following equality of Hodge numbers

$$h^{p,q}(X) = h^{n-p,q}(Y) \tag{14.32}$$

is true. In this case the pair (X, Y) will be called a mirror pair. Clearly with (X, Y) the pair (Y, X) will also be a mirror pair.

If we visualize the relation (14.32) in the Hodge diamonds of (strong) Calabi-Yau 3-manifolds (14.11) we see that the Hodge diamond of Y can be obtained from the Hodge diamond of X if we mirror it along the line passing trough the centre dividing the second and fourth quadrant. As the only nondetermined numbers in (14.11) are $h^{1,1}$ and $h^{2,1}$ the mirror conjecture for strong Calabi-Yau three-folds can be formulated such that to every Calabi-Yau three-fold X there exists a mirror Calabi-Yau three-fold Y such that

$$h^{1,1}(X) = h^{2,1}(Y) \quad \text{and} \quad h^{2,1}(X) = h^{1,1}(Y) . \tag{14.33}$$

I will explain in this section that the corresponding cohomology spaces $\mathrm{H}^1(X, \Omega_X)$ and $\mathrm{H}^1(X, \Omega_X^2)$ have geometric interpretations as infinitesimal deformations of Kähler structures compatible with the complex structure of X and infinitesimal deformations of complex structures, respectively. Hence, for mirror pairs (X, Y) in dimension 3 the moduli spaces of deformations of compatible Kähler structures on X and the moduli space of deformations of complex structures on Y (and vice versa) have the same dimensions.

But first it has to be pointed out that the mirror conjecture as formulated above cannot be true. To see this it has to be noted that there exist Calabi-Yau three-folds X with $h^{2,1}(X) = 0$, e.g. which are complex-analytically rigid, i.e. which cannot be deformed. For its conjectured mirror Y we obtain $h^{1,1}(Y) = 0$, in contradiction to the fact that the Kähler form of Y defines a non-vanishing class in $\mathrm{H}_{\bar{\partial}}^{1,1}(Y)$, see Proposition 13.23.

Hence a more reasonable form of the geometric conjecture is:

There exists classes of Calabi-Yau 3-manifolds whose members have mirror pairs such that (14.33) is true.

Of course a positive solution should exhibit these classes. We will approach this problem in the following. But first a reformulation of (14.33). Let (X, Y) be a mirror pair (of arbitrary dimension). Then by mirror symmetry and duality

$$H^r(X, \Omega_X^{n-s}) \cong H^{n-r}(Y, \Omega_Y^{n-s}) \cong H^r(Y, \Omega_Y^s) . \qquad (14.34)$$

By using the relation (14.6), valid for CY manifolds, the requirement of mirror symmetry can be described equivalently as

$$H^r(X, \bigwedge^s T_h M) \cong H^r(X, \Omega^s) . \qquad (14.35)$$

14.3.2 Batyrev's Results

In Sect. 14.4.1 an example of a CY mirror pair will be given. This example is the classical example of Candelas, de la Ossa, Green and Parkes [COGP]. Further examples of CY manifolds with mirrors are given as complete intersections (see their definition in Sect. 14.2).

Quite a number of further examples can be obtained as complete intersections in weighted projective space [CLS]. For the weighted projective space $\mathbb{P}(w_0, w_1, \ldots, w_n)$ the generating variables X_i have degree w_i with $gcd(w_0, w_1, \ldots, w_n) = 1$. Subvarieties are again zero sets of homogenous polynomials where we take into account that the monomial X^{d_i} should be considered to have degree $d_i w_i$. The usual projective space is given as $\mathbb{P}^n = \mathbb{P}(1, 1, \ldots, 1)$.

A very systematic approach incorporating all the above examples was given by Batyrev [Bat] by the class of hypersurfaces and more general complete intersections in toric varieties. An n-dimensional algebraic torus is the linear algebraic group $T := (\mathbb{C}^*)^n$ with group structure given by the component-wise multiplication in $\mathbb{C}^* := \mathbb{C} \setminus \{0\}$. A toric variety V is an algebraic variety (quasi-projective, projective, affine) with an action of an algebraic torus such that the action is "nearly transitive and free". This says that V has a Zariski open and dense orbit under the T action which is isomorphic to T.

Examples of toric varieties are $\mathbb{C}^n$ and $\mathbb{P}^n$. The torus action is given by the action of the diagonal matrices on the coordinates, i.e. by

$$(\lambda_1, \lambda_2, \ldots, \lambda_n) \quad \text{and} \quad \left(\left(\prod_{i=1}^n \lambda_i \right)^{-1}, \lambda_1, \ldots, \lambda_n \right) \qquad (14.36)$$

for $\mathbb{C}^n$ and $\mathbb{P}^n$, respectively. Typically toric varieties are constructed from convex polyhedra of certain types. The relevant numerical invariants (Euler number, Hodge numbers, etc.) of the toric varieties can be determined from the combinatorics of the polyhedra. Starting from a polyhedron its dual polyhedron can be defined.

Batyrev considers reflexive polyhedra. These are polyhedra such that the vertices have integer coordinates and the origin 0 lies inside of their interiors and the same holds for the dual polyhedra. If the original CY object is defined in the toric variety associated to the initial polyhedron Batyrev showed that the mirror object is defined in the toric variety associated to the dual polyhedron. Unfortunately, in general the objects will have singularities. These singularities have to be resolved to obtain manifolds in such a way that the canonical bundle stays trivial. This is only partially possible. But Batyrev showed that the singularities which cannot be resolved in this way lie in codimension ≥ 4. Hence for three-folds he indeed constructed a smooth CY mirror. In higher dimensions the mirror objects acquire singularities. Details can be found in [Bat] and [Voi].

14.3.3 Deformations

Let M be a compact complex manifold M of dimension n with holomorphic tangent bundle $T_h M$. Consider M for the moment as real differentiable manifold. The points of the moduli space (if it exists) of complex structures on the (differentiable) manifold M correspond to the set of complex structure of M modulo complex isomorphisms. In Chap. 7 we treated the situation of one-dimensional complex manifolds, i.e. the case of Riemann surfaces, in detail. The picture in higher dimensions is completely analogous.

Denote by $[M]$ the isomorphy class of the complex manifold M. Assume that the moduli space exists at least locally around the point $[M]$. Assume further that the moduli space is an analytic manifold. As in the one-dimensional case the tangent space of the moduli space at the point $[M]$ can be identified with the cohomology space $\mathrm{H}^1(M, T_h M)$. The elements of $\mathrm{H}^1(M, T_h M)$ correspond to infinitesimal deformations. By the assumed smoothness of the moduli space at $[M]$ the infinitesimal deformations can be lifted to local deformations around $[M]$. Other complex structures nearby can be obtained by deforming the given complex structure of M.

In the more general situation where smoothness is not guaranteed (but still assuming the existence of the moduli space) the obstruction for lifting infinitesimal deformations to local deformations lie in $\mathrm{H}^2(M, T_h M)$.

We come now back to our Calabi-Yau manifolds.

Theorem 14.11. (Bogomolov–Tian–Todorov–Ran). *Let X be a compact Kähler manifold with trivial canonical bundle K_X, then the moduli space of complex structures exists around $[X]$ and it is smooth.*

Hence $\dim \mathrm{H}^1(X, T_h X)$ gives the dimension of the moduli space at the point $[X]$ and every infinitesimal deformation can be lifted to a local deformation.

Let M be a CY three-fold. Using the relation (14.6) valid for Calabi-Yau manifolds we obtain

$$\mathrm{H}^1(M, T_h M) \cong \mathrm{H}^1(M, \Omega_M^2) \cong \mathrm{H}^2(M, \Omega_M^1) \, . \tag{14.37}$$

Hence the space of infinitesimal deformations of the complex structure of M is $h^{2,1}(M)$-dimensional. In particular, a CY three-fold M with $h^{2,1}(M) = 0$ is analytically rigid and vice versa.

Now we fix the complex structure but consider deformations of the Kähler structure compatible with the complex structure. We assume that M is a strong CY three-fold. In this case,

$$\mathrm{H}^2(M, \mathbb{C}) = \mathrm{H}^{1,1}_{\bar{\partial}}(M) = \mathrm{H}^{1,1}(M, \mathbb{C}) \,. \tag{14.38}$$

As already explained above, the set of Kähler structures on M is a cone in $\mathrm{H}^2(M, \mathbb{R})$, the real Kähler cone $K(M)$. From the fact that M is a strong CY three-fold the real cone $K(M)$ is open in $\mathrm{H}^2(M, \mathbb{R})$. Starting from a fixed Kähler form ω_0 we may add a form $\mu \in \mathrm{H}^{1,1}(M, \mathbb{C})$, which is a real, closed $(1,1)$-form. The sum $\omega = \omega_0 + \epsilon\mu$ will remain a real, closed $(1,1)$-form. Furthermore the form ω will stay positive definite if ϵ is small.

As we should describe everything in complex terms we need to consider the *complexified Kähler cone* $K_{\mathbb{C}}(M)$. We should allow to add forms in purely imaginary directions. It turns out that a convenient description of $K_{\mathbb{C}}(M)$ is given as the set

$$K_{\mathbb{C}}(M) := \{ [\omega] \in \mathrm{H}^{1,1}(M, \mathbb{C})/2\pi\mathrm{i}\,\mathrm{H}^2(M, \mathbb{Z}) \mid \mathbf{Re}(\omega) \text{ is a Kähler form} \} \,. \tag{14.39}$$

As $2\pi\mathrm{i}\,\mathrm{H}^2(M, \mathbb{Z})$ is a discrete lattice in $\mathrm{H}^{1,1}(M, \mathbb{C})$, locally the complexified Kähler cone at $[\omega_0]$ looks like $H^{1,1}(M, \mathbb{C})$ and the cone is an open subset. Indeed as everything is described explicitly in this way, the moduli space of Kähler deformations is a smooth manifold, in particular the deformations are unobstructed, i.e. the infinitesimal deformations can be lifted to local ones. Moreover, from the description it is clear that its complex dimension is $h^{1,1}(M)$.

So at least for mirror pairs we have the equality of the space of local deformations. This equality does not give any information about how the deformations are related. But mirror symmetry should also say something about a local identification of the moduli spaces. This identification does not necessarily take place at a fixed mirror pair, but it seems to be necessary to formulate it in terms of deformation families of the CY manifold. More precisely, it should give an identification in the neighbourhood of points at the boundary of the moduli spaces concerned. These boundary points are sometimes called *large Kähler limit points*, and *large complex limit points*, respectively. This also explains the fact that those CY manifolds with $h^{2,1} = 0$, i.e. without complex deformations, cannot have a mirror, at least not in the standard sense described.

Kontsevich's *homological mirror conjecture* aims also in this direction. It has to do with the isomorphy of two derived categories. The first derived category is the derived category of the bounded complex of coherent sheaves. As explained in Chap. 12 for affine varieties their geometry is encoded in the set of ideals of the coordinate rings. For general schemes (e.g. general algebraic manifolds) the globalization of ideals are coherent sheaves.

The corresponding derived category associated to the proposed mirror is the derived Fukaya A_∞-category related to certain Lagrangian submanifolds of the mirror. These are objects defined in terms of the symplectic geometry of the manifold. Kähler geometry is a special case of symplectic geometry. For a real differentiable manifold M a symplectic form ω is a closed nondegenerate 2-form. By the required nondegeneracity the real dimension of the manifold will always be even dimensional. Obviously, every Kähler form is also a symplectic form. A Lagrangian submanifold L of M is a submanifold of dimension $\dim M/2$ such that the restriction of the symplectic form vanishes.

A further aspect of mirror symmetry (related to Kontsevich's homological mirror conjecture) is the Strominger–Yau–Zaslov conjecture [SYZ]. It is based on open string theory and D-branes. It is an expected duality presented in terms of compatible fibrations by special Lagrangian tori.

14.4 Example of a Calabi-Yau Three-fold and Its Mirror: Results of Givental

The first examples of mirror pairs of Calabi-Yau three-folds were given by quintics in $\mathbb{P}^4$. For them the numerical mirror conjecture relating the generating function for the number of rational curves on X with the periods on the mirror family Y_λ was formulated in [COGP]. In the meantime Givental was able to prove it. First I will define the quintics and its mirror, then formulate the numerical mirror conjecture (now Givental's theorem) for them.

14.4.1 Quintics

Quintics are hypersurfaces of degree 5 in $\mathbb{P}^4$. Following the results in Sect. 14.2 about hypersurfaces, in particular by Proposition 14.10, the generic quintic X will be a strong CY three-fold, i.e. X will be connected and simply connected, with

$$K_X \cong \mathcal{O}_X, \quad \text{and} \quad h^{2,0}(X) = h^{0,2}(X) = 0 . \tag{14.40}$$

For the generic quintic we obtain as Euler characteristic

$$\chi(X) = -200 = 2(h^{1,1}(X) - h^{2,1}(X)) . \tag{14.41}$$

By the Lefschetz hyperplane theorem (see (14.24)) we obtain

$$h^{1,1}(X) = h^{1,1}(\mathbb{P}^n) = 1 . \tag{14.42}$$

Hence

$$h^{2,1}(X) = 101, \quad h^{1,1} = 1 . \tag{14.43}$$

A special case is given by the Fermat quintic X_F defined as the zero set of the polynomial

$$X_0^5 + X_1^5 + X_2^5 + X_3^5 + X_4^5 . \tag{14.44}$$

To construct the mirror of the quintics we start from X_F. Let $\mathbb{Z}/5\mathbb{Z}$ be the group of fifth roots of units. The group $(\mathbb{Z}/5\mathbb{Z})^5$ acts on the points of $\mathbb{P}^4$ by the action

$$(\zeta_0, \zeta_1, \ldots, \zeta_4).(X_0 : X_1 : \cdots : X_4) := (\zeta_0 X_0 : \zeta_1 X_1 : \ldots : \zeta_4 X_4) \qquad (14.45)$$

on the homogenous coordinates. The diagonally embedded copy $\mathbb{Z}/5\mathbb{Z}$ acts trivially on $\mathbb{P}^4(\mathbb{C})$. Hence the group effectively acting is $(\mathbb{Z}/5\mathbb{Z})^5/diag(\mathbb{Z}/5\mathbb{Z})$. Under this action the points on X_F are mapped to points of X_F, i.e. X_F is an invariant hypersurface.

Let G be the subgroup defined by the elements fulfilling $\prod_{i=0}^4 \zeta_i = 1$. We take the quotient X_F/G. This quotient has singularities. But they can be resolved in such a way that the desingularization Y has trivial canonical bundle and is again a Calabi-Yau three-fold. One can show

$$h^{2,1}(Y) = 1, \qquad h^{1,1}(Y) = 101 . \qquad (14.46)$$

If we compare this with (14.43) we see that Y is indeed a geometric mirror for every quintic in $\mathbb{P}^4$.

As already explained above, mirror symmetry is a relation which has to do with deformation families. The moduli space of complex deformations of a quintic X is 101 dimensional (as $h^{2,1}(X) = 101$). Quintics are given by homogeneous polynomials of degree 5 in five variables. The space of quintic polynomials is $\binom{5+4}{4} = 126$ dimensional. Two quintics are isomorphic if they are identified under a projective coordinate transformation. The corresponding group is GL(5) which is 25 dimensional. Hence the space of isomorphy classes of quintics is 101 dimensional. In particular for a generic quintic its nearby deformations are again smooth quintics.

The one parameter deformation of the mirror can easily be found. Instead of the Fermat quintic we consider the deformed quintic X_λ defined by

$$X_0^5 + X_1^5 + X_2^5 + X_3^5 + X_4^5 + \lambda X_0 X_1 X_2 X_3 X_4 . \qquad (14.47)$$

Again the group G operates on the deformed X_λ. We take its quotient, desingularize it and obtain a one-dimensional family Y_λ of complex deformations of Y as clearly $Y_0 = Y$.

The large complex limit point of the mirror family is given by "$\lambda = \infty$". Before taking the limit we should divide (14.47) by λ. As long as $\lambda \neq 0$ the corresponding hypersurface will be the same as the one defined by (14.47). But for $\lambda = \infty$ only the term $X_0 X_1 X_2 X_3 X_4$ will remain. The corresponding hypersurface will be a union of linear hyperplanes with singularities at their intersections. In particular, already before the group action the manifold will be singular.

14.4.2 The Counting of Rational Curves

For the quintic and its mirror we are able to write down the famous relation between the generating function for the number of rational curves on X and certain period functions for the mirror family Y_λ. This example is due to [COGP]. I follow closely the presentation of Kontsevich [Ko1], see also [Pan], [Gi].

First we have to define what we mean by smooth rational curve of degree d on X. Of course C should be a complex one-dimensional submanifold which is isomorphic to $\mathbb{P}^1$. As such it determines a class in $H_2(X, \mathbb{C})$. The space $H^2(X, \mathbb{C})$ is dual to $H_2(X, \mathbb{C})$. As in our case $H^2(X, \mathbb{C}) = H^{1,1}(X, \mathbb{C})$, it is generated by the Kähler form ω obtained by restricting the Kähler form of the projective space to the quintic. The degree of the curve C is defined to be the class evaluated against the generator of $H^{1,1}(X, \mathbb{C})$, i.e. the Kähler form integrated along the curve. In our context of a hypersurface X in $\mathbb{P}^4$ this can be described in more geometric terms as follows. If C is a curve on X, its degree will be given as the number of intersection points between C and a generic hyperplane.

Let n_d be the number of smooth rational curves of degree d on X (a generic quintic). For the following formulas we have to assume that these curves lie isolated, i.e. there are no continuous families of such curves. Recall that for the involved models in physics the maps from the world-sheet into the CY manifold is of importance. Hence we have to count maps from $\mathbb{P}^1 \to X$. Any such map can be realized as a covering map $\mathbb{P}^1 \to \mathbb{P}^1$ of degree k and an embedding of $\mathbb{P}^1$ into X. To take this into account one considers the modified numbers

$$N_d = \sum_{k|d} \frac{1}{k^3} n_{d/k} . \tag{14.48}$$

Between these numbers we have the relation

$$\sum_{d=1}^{\infty} N_d q^d = \sum_{d=1}^{\infty} \sum_{k=1}^{\infty} \frac{1}{k^3} n_d q^{k \cdot d} . \tag{14.49}$$

The numbers N_d are the genus 0 Gromov–Witten invariants of the quintic.

As generating function one takes

$$F(t) := \frac{5}{6} t^3 + \sum_{d \geq 1} N_d \exp(d \cdot t) . \tag{14.50}$$

Before we discuss its relation to the period function some remarks on this formulas. First, not for all quintics the rational curves lie isolated. The Fermat quintic is an example for which this is not the case. For it the set of degree one rational curves form a continuous family. Clemens' conjecture says that at least on a generic quintic they lie isolated. Only, very recently (2005) Wang announced a proof of Clemens' conjecture [Wa]. In [Ko1] Kontsevich gives a more conceptional description for the (virtual) number of curves which works also in the cases for which the curves are not isolated, resp. when they are singular. This formula is based on the Gromov–Witten invariants N_d and quantum cohomology. They are always defined and via (14.49) the (virtual) numbers of rational curves even if they do not exist in the strict sense. It turned out that this description via quantum cohomology is the right way to approach the problem.

For our mirror family Y_λ we make the substitution for the parameter $\lambda = -z^{-1/5}$, resp. $z = -\lambda^{-5}$. We will still denote the family by Y_z. In particular our large complex limit point will be reached by $z = 0$.

For the mirror family the following functions (depending on the parameter z) are defined:

$$\psi_0(z) = \sum_{n=0}^{\infty} \frac{(5n)!}{(n!)^5} z^n ,$$

$$\psi_1(z) = \log z \cdot \psi_0(z) + 5 \sum_{n=1}^{\infty} \frac{(5n)!}{(n!)^5} \left(\sum_{k=n+1}^{5n} \frac{1}{k} \right) z^n ,$$

$$\psi_2 = \frac{1}{2} (\log z)^2 \cdot \psi_0 + \cdots ,$$

$$\psi_3 = \frac{1}{6} (\log z)^3 \cdot \psi_0 + \cdots .$$

(14.51)

The functions $\psi_i(z)$ are periods $\int_{\gamma_i} \omega$ of the Calabi-Yau manifold Y_z. They can be given as the solutions of the following linear differential equation of 4th order

$$\left(\left(z \frac{d}{dz} \right)^4 - 5z \left(5z \frac{d}{dz} + 1 \right) \left(5z \frac{d}{dz} + 2 \right) \left(5z \frac{d}{dz} + 3 \right) \left(5z \frac{d}{dz} + 4 \right) \right) \psi(z) = 0 ,$$

(14.52)

given by

$$\sum_{i=0}^{3} \psi_i(z) \epsilon^i + O(\epsilon^4) = \sum_{n=0}^{\infty} \frac{(1+5\epsilon)(2+5\epsilon) \cdots (5n+5\epsilon)}{((1+\epsilon)(2+\epsilon) \cdots (n+\epsilon))^5} z^{n+\epsilon} .$$

(14.53)

We introduce the new variable

$$t(z) = \frac{\psi_1(z)}{\psi_0(z)} ,$$

(14.54)

then the famous mirror relation is

$$F \left(\frac{\psi_1}{\psi_0} \right) = \frac{5}{2} \cdot \frac{\psi_1 \psi_2 - \psi_0 \psi_3}{\psi_0^2} .$$

(14.55)

Hence at least in principle, if we use in (14.50) as t the new variable (14.54) and compare it with (14.55) then the numbers N_d and n_d of rational curves can be expressed in terms of the expansions of the ψ_i. See also [Pan] and [Gi] for a slightly different but equivalent presentation of the involved objects.

The formula (14.55) was proved by Givental [Gi]. More precisely, he deduced the general form relating the generating function for the (virtual) number of rational curves on X with certain hypergeometric functions related to the mirror Y for arbitrary CY hypersurfaces in projective space. This has been extended by him to CY complete intersections in smooth toric varieties. The use of Gromov–Witten invariants and equivariant cohomology is important. For the proof I have to refer to the original work [Gi] and to the Bourbaki exposé of Pandharipande [Pan].

For small d the numbers are

$$n_1 = 2875, \quad n_2 = 6,092,550, \quad n_3 = 317,206,375, \quad n_4 = 242,467,530,000\,.$$
$$(14.56)$$

The number n_1 gives the number of straight lines on the quintic and it is classical. The numbers n_2 and n_3 have been calculated by Katz, and Ellingsrud and Stromme, respectively by algebraic-geometric methods. The number n_4 was conjectured in [COGP] on the basis of the numerical mirror conjecture. It was Kontsevich who proved it via the theory of torus localization. See [Pan] for further details and the necessary references.

Physicists speak in this context often from Yukawa coupling functions. These functions are correlation functions of the considered σ-models. In mathematical terms (see [Voi], [Mor]) they can be given as certain cubic forms on the tangent spaces of the families of complex deformations and Kähler deformations, i.e. on $H^1(X, T_h X)$ and on $H^1(X, \Omega_X)$, respectively. As formulated in (14.35) mirror symmetry says that these two spaces are isomorphic. Physicists conjectured that both Yukawa coupling functions can be related in a precise sense. In the quintic example one of the function is essentially given by a modification of the function (14.50). The other is expressed in terms of the period functions. Equation (14.55) is the proposed relation.

Hints for Further Reading

General mathematical background information can be found in

[BPV]   Barth, W., Peters, C., Van de Ven, A., *Compact Complex Surfaces*, Springer, 1984.

[GH]   Griffiths, Ph., Harris, J., *Principles of Algebraic Geometry*, Wiley, New York, 1978.

[We]   Wells, R.O., *Differential Analysis on Complex Manifolds*, Springer, 1980.

The following references are review articles and books on different aspects of mirror symmetry:

[CK]   Cox, D., Katz, S., Mirror Symmetry and Algebraic Geometry, *Math. Surv. Monogr.*, **68**, AMS, Providence, RI, 1999.

[GHJ]   Gross, M., Huybrechts, D., Joyce, D., *Calabi-Yau Manifolds and Related Geometries*, Universitext, Springer, 2003.

[Ko1]   Kontsevich, M., Homological Algebra of Mirror Symmetry, (in) *Proc. of the ICM Zürich 1994*, 120–139, Birkhäuser, Basel, 1994.

[KO]   Kapustin, A.N., Orlov, D.O., Lectures on Mirror Symmetry, Derived Categories, and D-Branes, *Russ. Math. Surv.* **59:5** (2004) 907–940.

[Ma1]   Manin, Y.I., *Moduli, Motives, Mirrors*, math.AG/0005144.

[Mor]   Morrison, D., Mirror Symmetry and Rational Curves on Quintic Threefolds. A Guide for Mathematicians, *J. AMS* **6** (1993), 223–247.

[Mor2] Morrison, D., Geometric aspects of mirror symmetry, (in) *Mathematics Unlimited – 2001 and Beyond*, B. Engquist, W. Schmid (eds), Springer, 2001.

[Pan]   Pandharipande, R., Rational Curves on Hypersurfaces (After A. Givental), Séminaire Bourbaki, Exp. 848, *Astérisque* **252** (1998) 307–340, math.AG/9806133.

[Th]   Thomas, R.P., The Geometry of Mirror Symmetry, math.AG/0512412.

[Voi]   Voisin, C., *Mirror Symmetry*, SMF/AMS Texts and Monographs, **1**, AMS, Providence, RI, 1999.

The following references are some original research articles quoted in this section. For more references you should consult the above review articles.

[Bat]   Batyrev, V.V., Dual Polyhedra and Mirror Symmetry for Calabi-Yau Hypersurfaces in Toric Varieties, *J. Algeb. Geom.* **3** (1994) 493–535.

[CLS]   Candelas, P., Lynker, M., Schimmrigk, R., Calabi-Yau Manifolds in Weighted $\mathbb{P}_4$, *Nucl. Phys.* **B 341** (1990) 384–402.

[COGP] Candelas, P., de la Ossa, X., Green, P., Parkes, L., A Pair of Calabi-Yau Manifolds as an Exactly Soluble Superconformal Theory, *Nucl. Phys.* **B 359** (1991) 21–74.

[Gi]   Givental, A.B., Equivariant Gromov–Witten Invariants, *IMRN* **13** (1996) 613–663.

[LLY]   Lian, B.H., Liu, K., Yau, S.-H., Mirror Principle. I. *Asian J. Math.* **1** (1997) 729–763, math.AG/9712011.

[SYZ] Strominger, A., Yau, S.T., Zaslov, E., Mirror Symmetry Is T-Duality, *Nucl. Phys.* **B 479** (1996) 243–259.

[Wa]   Wang, B., *Clemens's conjecture. I.*, math.AG/0511312; *Clemens's conjecture. II.*, math.AG/0511364.

Appendix

p-adic Numbers

In Chap. 7 we saw that the moduli space of curves over the complex numbers comes in a natural manner with an arithmetic structure. Just to remind you let me repeat the following facts. If $\mathcal{C}$ is a curve, then it is the set of common zeros of finitely many homogeneous polynomials. For example, a curve in the projective plane can be given by the zeros of one homogeneous polynomial. If after a suitable change of coordinates the polynomials can be given in such a way that all coefficients are rational numbers then we say that the curve $\mathcal{C}$ can be *defined over the rational numbers* $\mathbb{Q}$.

If g is a rational polynomial of degree k we can write it as

$$g = \sum_i r_i m_i, \qquad r_i = \frac{s_i}{t_i}$$

where m_i are monomials (just products of the variables) and the s_i and t_i are integers. If we now multiply g by the least common multiple t of the t_i we get

$$\hat{g} = t \cdot g = \sum_i \hat{s}_i \cdot m_i$$

with $\hat{s}_i$ integers. Of course g and $\hat{g}$ have the same set of points as zeros. Hence every curve defined over the rational numbers $\mathbb{Q}$ can be defined over the integers $\mathbb{Z}$. Hence we can use all number theoretic techniques to get information on the geometric (and arithmetic) structure. (This works also in the opposite direction.)

Now, how do we get from the rational numbers to the complex numbers? Of course this is well known. But the technique is also essential in the case of the *p*-adic numbers, so let me repeat it briefly. For the rational numbers we have the so-called *absolute value*. It allows us to define a topology and the notion of a convergent sequence. As you know there exist sequences of rational numbers which "converge" but have no limits in $\mathbb{Q}$ (of course strictly speaking we do not talk about convergence in this case). We enlarge $\mathbb{Q}$ by adding these limit points to get the real numbers $\mathbb{R}$.

M. Schlichenmaier, *p*-adic Numbers. In: M. Schlichenmaier, An Introduction to Riemann Surfaces, Algebraic Curves and Moduli Spaces, Theoretical and Mathematical Physics, 203–212 (2007)
DOI 10.1007/978-3-540-71175-9_16

One way to construct $\mathbb{R}$ in more precise terms is the following. We call a sequence (x_n) a Cauchy sequence if for every rational number $\epsilon > 0$ there exists a natural number n_0 such that

$$|x_n - x_m| < \epsilon \qquad \text{if } n, m > n_0.$$

As remarked above not every Cauchy sequence has to have a limit. But if it has a limit and this limit is 0, then we call (x_n) a *zero sequence*.

We can define addition, subtraction and multiplication on the set of Cauchy sequences element-wise, for example

$$(x_n) + (y_n) := (x_n + y_n).$$

The real numbers are now defined as the set of classes of Cauchy sequences modulo zero sequences. This means that we identify two Cauchy sequences (x_n) and (z_n) if $(x_n) - (z_n)$ is a zero sequence. All operations can be well defined on these classes. In addition if $[(x_n)]$ is not the zero class (equivalently (x_n) is not a zero sequence) we can define the multiplicative inverse $[(x_n)]^{-1}$ in the following way. By adding a suitable zero sequence to (x_n) we can arrange that for the new sequence (still denoted by (x_n)) $x_n \neq 0$ for all n. In this way we stay in the same class. We set

$$[(x_n)]^{-1} := \left[\left(\frac{1}{x_n} \right) \right]$$

which is again a Cauchy sequence, because the x_n are bounded away from 0.

It is easy to see that $\mathbb{R}$ will be a field and that $\mathbb{Q}$ is embedded as a subfield via the constant sequence

$$[(x_n)] \qquad \text{for all } n : x_n = x \in \mathbb{Q}.$$

Now we can also define the absolute value in $\mathbb{R}$ by setting

$$|[(x_n)]| := [(|x_n|)].$$

With this we see that in $\mathbb{R}$ all Cauchy sequences have a limit. Hence we call $\mathbb{R}$ the completion of $\mathbb{Q}$ with respect to the absolute value $|.|$.

From the mathematical viewpoint $\mathbb{R}$ is not yet fully satisfying: we cannot solve every algebraic equation. In technical terms $\mathbb{R}$ is not *algebraically closed*. For this reason we have to "add" a single element, the root i of the polynomial equation

$$x^2 + 1 = 0.$$

It obeys the multiplicative law $i^2 = -1$. In this manner we get the complex numbers $\mathbb{C}$. We can extend the absolute value in $\mathbb{R}$ to the usual complex absolute value. $\mathbb{C}$ is now an algebraically closed complete field.

But what forces us to stick to this special absolute value to start from the rational numbers? Let us list the characteristic features of the absolute value which allowed us to work as above. The absolute value is a map

$$|\cdot| : \mathbb{Q} \to \mathbb{Q}$$

with

$$|a| \geq 0, \quad a \in \mathbb{Q}, \qquad |a| = 0 \quad \text{if and only if } a = 0. \tag{1}$$
$$|a \cdot b| = |a| \cdot |b|, \quad a, b \in \mathbb{Q} \quad \text{(multiplicative law)} \tag{2}$$
$$|a + b| \leq |a| + |b|, \quad a, b \in \mathbb{Q} \quad \text{(triangle identity)} \tag{3}$$

If we have such a map we call it a *valuation* for the field $\mathbb{Q}$. We can generalize this immediately to arbitrary fields (for the values we remain in $\mathbb{Q}$ or $\mathbb{R}$).

There is one map, the so-called *trivial valuation*, defined by

$$|a| = \begin{cases} 1, & \text{if } a \neq 0 \\ 0, & \text{if } a = 0 . \end{cases}$$

This valuation is of no interest to us. If we use the word valuation we always assume a nontrivial one.

Let us consider a very interesting one. If

$$r = \frac{n}{m} \in \mathbb{Q}, \quad r \neq 0, \quad n, m \in \mathbb{Z}$$

then we can write

$$n = 2^s n', \quad m = 2^t m'$$

where n' and m' are integers which are not divisible by 2. We set

$$|r|_2 := 2^{t-s} \text{ and } |0|_2 := 0.$$

Let us check the conditions above.
Condition 1: is clear.
Condition 2: Normalized as above we get

$$r_1 = \frac{n_1}{m_1} = 2^l \cdot \frac{n_1'}{m_1'}, \quad r_2 = \frac{n_2}{m_2} = 2^p \cdot \frac{n_2'}{m_2'},$$
$$|r_1|_2 \cdot |r_2|_2 = 2^{-l} 2^{-p} = 2^{-(l+p)}$$
$$r_1 r_2 = 2^{l+p} \cdot \frac{n_1' n_2'}{m_1' m_2'}$$

Because 2 is a prime number it divides neither $n_1' n_2'$ nor $m_1' m_2'$. Hence we get

$$|r_1 r_2|_2 = 2^{-(l+p)}.$$

This is what we had to show.

Condition 3: We keep the same notation as above. We separate two cases.

($l \neq p$): without restriction we assume $l < p$.

$$r_1 + r_2 = 2^l \cdot \left(\frac{n_1'}{m_1'} + 2^{p-l} \frac{n_2'}{m_2'} \right) = 2^l \cdot \frac{n_1' m_2' + 2^{p-l} n_2' m_1'}{m_1' m_2'}.$$

Now 2 divides neither the numerator (because 2 divides only one term in the sum) nor the denominator, hence

$$|r_1 + r_2|_2 = 2^{-l} = |r_1|_2 \leq |r_1|_2 + |r_2|_2.$$

($l = p$): here we get

$$r_1 + r_2 = 2^l \cdot \frac{n_1' m_2' + n_2' m_1'}{m_1' m_2'}.$$

It can happen that 2 divides the numerator, hence

$$|r_1 + r_2|_2 = 2^{-(l+q)} \leq |r_1|_2 \leq |r_1|_2 + |r_2|_2$$

with q a positive integer.

In both cases condition 3 is fulfilled. Even more is true. We can replace (3) by

$$|r_1 + r_2| \leq \max(|r_1|, |r_2|) \tag{3'}$$

and we know in addition

$$|r_1 + r_2| = \max(|r_1|, |r_2|) \quad \text{if } |r_1| \neq |r_2|$$

If the stronger condition (3') is valid we call $|\,.\,|$ a *nonarchimedian* valuation. Another name for it is an *ultrametric*. In a nonarchimedian valuation the integers are always within the "unit circle". We can see this as follows. We have $|1| = 1$ as usual ($|a| = |1 \cdot a| = |1| \cdot |a|$). But now

$$|2| = |1 + 1| \leq \max(|1|, |1|) = |1| = 1.$$

By induction we get $|n| \leq 1$ for all integers n. This is a rather strange result in contrast to the usual absolute value.

In the above the only fact we used about the number 2 was that it was a prime number. Hence we can define in an identical manner

$$|a|_p = p^{-s} \quad \text{if } a = p^s \cdot \frac{n'}{m'}, \quad |0|_p = 0$$

where p divides neither n' nor m'. We call $|\,.\,|_p$ the *p-adic valuation*. If there is a danger of confusion we use $|\,.\,|_\infty$ for the usual absolute value.

We can define a topology, convergence, Cauchy sequences and so on. With this we see that all these valuations are essentially different. To see this let us consider the sequence

$$(x_n) = (p^n)$$

for a fixed prime number p. Now

$$|p^n|_\infty = p^n$$

and hence it is unbounded and diverges in the usual topology. On the other hand

$$|p^n|_p = p^{-n}, \qquad |p^n|_q = 1, \quad \text{for } q \neq p.$$

We see (x_n) is a zero sequence in the p-adic topology and a bounded (but not zero) sequence in the q-adic topology.

We call two valuations $|\cdot|_a, |\cdot|_b$ equivalent if we have

$$|\cdot|_a = |\cdot|_b^r$$

with a positive real number r, i.e. if they define the same topology.

Theorem A.1. *(Ostrowski)* $|\cdot|_\infty$ *and* $|\cdot|_p$, *(p all prime numbers), are representatives of all equivalence classes of valuations for the rational number field* $\mathbb{Q}$.

In the same way as we constructed the real numbers $\mathbb{R}$ via Cauchy sequences with respect to the absolute value we construct the *p-adic numbers* $\mathbb{Q}_p$ starting with the p-adic valuation for every prime number p. Our field $\mathbb{Q}$ lies in all these completions:

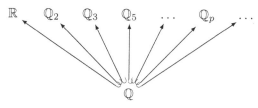

In $\mathbb{Q}_p$ we can do analysis similar to real analysis. Just as we can represent every real number as a decimal expansion, we can represent every p-adic number by a p-adic expansion. Take $a \in \mathbb{Q}_p$, by extension of the p-adic valuation to $\mathbb{Q}_p$ we calculate

$$|a|_p = p^{-m}, \quad m \in \mathbb{Z}.$$

With this we can write

$$a = \sum_{k \geq m}^{\infty} a_k p^k = \lim_{n \to \infty} \sum_{k \geq m}^{n} a_k p^k$$

with

$$a_k \in \{0, 1, 2, \ldots, p-1\}.$$

As you can see the sum goes in the opposite direction of the decimal expansion.

In some sense the analysis in $\mathbb{Q}^p$ is much easier than that in $\mathbb{R}$ but more unfamiliar. For example we have in $\mathbb{Q}^p$ the following:

Lemma A.2. *A series $\sum_k a_k$ converges if and only if the sequence (a_k) is a zero sequence.*

(Quite a number of students starting to learn mathematics would be happy if this would work also in $\mathbb{R}$.)

Proof. Let (a_k) be a zero sequence. Then $(m \geq n)$

$$\left| \sum_{k=1}^{m} a_k - \sum_{k=1}^{n} a_k \right| = \left| \sum_{k=n}^{m} a_k \right| \leq \max_{k=n}^{m} (|a_k|).$$

Because (a_k) is a zero sequence the right-hand side can be made as small as one wants by choosing n big enough. This means that the sequence of partial sums is a Cauchy sequence and hence has a limit. The other direction is the usual argument of real analysis. $\square$

Now we can also define p-adic power series like the *exponential*

$$E(x) := \sum_{k=0}^{\infty} \frac{x^k}{k!}$$

and the *logarithm*

$$\log(1+x) = \sum_{k=1}^{\infty} (-1)^{k+1} \frac{x^k}{k}.$$

Despite the fact that the expressions are the same you might expect that the domain of convergence will be different. We write

$$|x|_p = p^{-\operatorname{ord}_p(x)}$$

to define the *p-adic order*. With this we can calculate that $E(x)$ is convergent if and only if $\operatorname{ord}_p(x) > \frac{1}{p-1}$ and $\log(1+x)$ is convergent if and only if $\operatorname{ord}_p(x) > 0$.

This is again a rather strange behaviour. If $p \neq 2$ and x is an integer then $E(x)$ is a convergent series if and only if x is a multiple of p. If $p = 2$ and everything else stays the same, then x has to be a multiple of 4.

One can now introduce p-adic measures, distributions, integrals and Fourier transformations.[1] Let me mention that in $\mathbb{Q}_p$ there is in contrast to $\mathbb{R}$ still a lot of number theory involved. For example, we have also the *p-adic integers* which are defined to be the $x \in \mathbb{Q}_p$ with $\operatorname{ord}_p(x) \geq 0$. Like in the case of the rational numbers we can write every element of $\mathbb{Q}_p$ as quotient of two p-adic integers.

We are still not at the end in our construction of the analogue of $\mathbb{C}$. To get from $\mathbb{R}$ to $\mathbb{C}$ we took the algebraic closure of $\mathbb{R}$. This can also be done with $\mathbb{Q}_p$. In this case we have to "add" infinitely many roots of polynomials to get

---

[1] See Koblitz, [Ko], p. 30.

to $\mathbb{Q}_p^{a.c.}$. Of course this $\mathbb{Q}_p^{a.c}$ also admits an extension of the p-adic valuation. But in contrast to $\mathbb{C}$ it is no longer complete. We have to again construct the completion of this field. The resulting field is algebraically closed and complete.

Of course, one might ask why one should bother about the p-adic numbers if one is not interested in number theory. Some idea of their importance outside of number theory might be given by the following product formula for all rational numbers $x \neq 0$:

$$|x|_\infty \cdot \prod_{p \in \mathbb{P}} |x|_p = 1.$$

(We use $\mathbb{P}$ to denote the set of all prime numbers.) It says, if we know all p-adic values of a rational number we also know its absolute value. This is a reformulation of the trivial fact that we know the absolute value of a rational number if we know how often all primes appear in the numerator and in the denominator. But there is more behind it. The mathematical idea is that the rational numbers are the objects of primary interest. But they are very difficult to handle from the arithmetic viewpoint. $\mathbb{R}$ and all $\mathbb{Q}_p$ are much easier. Each of them reflects one facet of the complexity of $\mathbb{Q}$. From this viewpoint one calls $\mathbb{Q}$ a *global* field and $\mathbb{R}$ and the different $\mathbb{Q}_p$ *local* fields. If $x \in \mathbb{Q}$ is a number fulfilling a suitable relation (for example to be a zero of a polynomial equation with integral coefficients) then x considered as element in $\mathbb{R}$ and $\mathbb{Q}_p$ clearly fulfils the transformed relation in $\mathbb{R}$, resp. in $\mathbb{Q}_p$. The converse problem is the one that really matters. If we can solve the transformed relations by elements x_∞ and x_p, $p \in \mathbb{P}$ (which in most cases is easier) we can ask under which conditions we can assemble this information to get a solution for the starting relation in $\mathbb{Q}$. (This is also known under the name *local–global principle*.)

In the above you might get the (right) feeling that it is better to consider the real number field and all p-adic number fields simultaneously. This is formalized in the concept of *adèles*. We start with the infinite product

$$\mathbb{R} \times \prod_{p \in \mathbb{P}} \mathbb{Q}_p.$$

It consists of all infinite sequences

$$(x_\infty, x_2, x_3, x_5, \ldots, x_p, \ldots)$$

where $x_\infty \in \mathbb{R}, x_p \in \mathbb{Q}_p$. Now we consider the subset with

$$|x|_p \leq 1$$

for almost every p (which is shorthand for: the condition can be violated only by a finite number of primes). This subset is called the set of *adèles* and is denoted by $\mathbb{A}_\mathbb{Q}$. It is a ring if we define addition and multiplication componentwise. The set of multiplicative invertible elements in $\mathbb{A}_\mathbb{Q}$ is called *idèles*. $\mathbb{Q}$ is now embedded diagonally into $\mathbb{A}_\mathbb{Q}$ by

$$x \mapsto (x, x, x, \ldots).$$

At first sight this is only a formal tool. But because this is a ring we can do algebraic geometry with it and other nice things. The following is an example taken from Manin (details can be found there [Man]). One can define the group $\mathrm{Sl}(2, \mathbb{A}_\mathbb{Q})$ in a suitable manner. By diagonal embedding, $\mathrm{Sl}(2, \mathbb{Q})$ is a discrete subgroup. Now the factor group is a compact group on which we can integrate. By normalizing the measure and splitting it up into the factors corresponding to factor groups of $\mathrm{Sl}(2, \mathbb{R})$ and $\mathrm{Sl}(2, \mathbb{Q}_p)$ we get the result

$$1 = \frac{\pi^2}{6} \cdot \prod_{p \in \mathbb{P}} (1 - p^{-2}).$$

This relates the arithmetically defined product expression with the transcendental number π. The product is an object which is defined by knowing the measure on the p-adic groups and it determines the analogous measure on the real group. Of course the above formula is not a new result gained only by the use of adèles. But the adèles gives a new insight. Classically it comes from the *Riemann zeta function*

$$\zeta(s) = \sum_{n \geq 1} \frac{1}{n^s}$$

where we have

$$\zeta(2) = \frac{\pi^2}{6} = \prod_{p \in \mathbb{P}} (1 - p^{-2})^{-1}$$

which is the above relation.

Up to now we have done everything over the rational numbers. In fact this is not sufficient. We have to allow finite algebraic field extensions of $\mathbb{Q}$. We can obtain these in the following way. We embed $\mathbb{Q}$ in $\mathbb{C}$ and take an $\alpha \in \mathbb{C}$ which is a root of a rational irreducible polynomial f. (Here irreducible means it does not have a factorization into two nonconstant rational polynomials.) The imaginary unit i for example is such an α. It is the zero of the polynomial

$$\mathbf{X}^2 + 1.$$

Another example is

$$\alpha = \zeta_3 = \exp\left(\frac{2\pi i}{3}\right),$$

a so-called 3rd root of unity. It is a zero of the polynomial

$$\mathbf{X}^2 + \mathbf{X} + 1.$$

We denote the smallest subfield of $\mathbb{C}$ containing $\mathbb{Q}$ and α by $\mathbb{Q}(\alpha)$. It is also a finite-dimensional $\mathbb{Q}$-vector space. Its basis is given by

$$1, \alpha, \alpha^2, \ldots, \alpha^m, \quad m = \deg f - 1.$$

Such a field is called a number field. Just as we can build $\mathbb{Q}$ out of the integers as their quotients, we can find "integers" in $\mathbb{Q}(\alpha)$ which do the job here. In contrast to the usual integers , the *rational integers*, these are called *algebraic integers*. With these algebraic integers we can do arithmetic, define divisibility and so on. Unfortunately we have in general no unique factorization of every algebraic integer into primes. Even worse, we have to drop the notion of prime number at all. But there is a rather useful substitute for the prime numbers. The prime numbers of $\mathbb{Z}$ are in a 1:1 correspondence with the classes of nonarchimedian valuations of $\mathbb{Q}$. Now we ask for such nonarchimedian valuations for $\mathbb{Q}(\alpha)$. Let $|\,.\,|$ be such a valuation. By restricting it to $\mathbb{Q}$ we get one of the p-adic valuations. Conversely one can show that for every p-adic valuation on $\mathbb{Q}$, there exist only finitely many extensions to the whole field $\mathbb{Q}(\alpha)$. By embedding into $\mathbb{C}$ the absolute value is already extended to the usual complex absolute value on $\mathbb{Q}(\alpha)$. Different extensions of the absolute value correspond to different embeddings. There are also finitely many of them.

Now we can do everything (completion, adèles, ...) we did for $\mathbb{Q}$ for this number field $\mathbb{Q}(\alpha)$. The corresponding complete field is a more tractable subfield of the big p-adic algebraically closed and complete field.

Hints for Further Reading

Further details can be found in

[Ko] Koblitz, N., *p-adic Numbers, p-adic Analysis and Zeta Functions*, Springer, 1977.

It contains the basic concepts including p-adic integration.
Of course you also find the basics in other books like

[Ba] Bachmann, G., *Introduction to p-adic Numbers and Valuation Theory*, Academic Press, New York 1964.

[Wa] Van der Waerden, B.L., *Algebra II*, Springer, 1967.

Adèles can be found in

[Ta] Tamagawa, T., Adèles, *Proc. Symp. Pure Math.* **9** (1966) 113–121, Amer. Math. Soc., Providence.

[Wei] Weil, A., *Basic Number Theory (3rd rev.ed.)*, Grundl. Math. Wiss. 144, Springer, 1974.

Concerning the more speculative aspects of application of p-adic numbers to string theory, see

[Man] Manin, Y.I., Reflections on Arithmetical Physics, Talk at Poiana-Brasov School on Strings and Conformal Field Theory, 1–14. September. 1987, (appeared in) *Perspect. Phys.*, Academic Press, Boston, MA, 1989.

Since the first edition of this book, p-adic and adelic mathematical physics developed further in the context of p-adic string theory and cosmology. For these more recent developments see

[BF]   Brekke, L., Freund, P.G.O., p-adic Numbers in Physics, *Phys. Rep.* **233** (1993) 1–66.

[DDNV] Djordjević, G.S., Dragovich, B., Nesić, L., Volovich, I.V., p-adic and Adelic Minisuperspace Quantum Cosmology, *Int. J. Mod. Phys.*, A **17** (2002) 1413–1433.

[KRV] Khrennikov, A.Y., Racić, Z., Volovich I.V. (eds), *p-adic Mathematical Physics, 2nd International Conference*, Belgrade, Serbia, and Montenegro, 15–21.9.2005, AIP, Melville, New York, 2006.

Index

Abel's theorem, 58
abelian varieties, 69
absolute value, 203
adèles, 209
adelic string theory, 84
adjunction formula, 189
affine scheme, 160
affine scheme associated to a
    ring, 156
affine space, 134
affine variety, 95
algebraic bundles, 91
algebraic genus, 71
algebraic map, 64
algebraic set, 134
algebraic space, 167
algebraic stack, 167
algebraic variety, 63
analytic isomorphism, 14
analyticmap, 14
annulator, 151
annulator ideal, 147
arithmetic structure, 85
associated sheaf, 119
atlas, 7

Batyrev's results, 193
Belavin-Knizhnik theorem, 123
Betti number, 24, 176
biholomorphic map, 14
Bogomolov-Tian-Todorov-Ran theorem,
    194
boundaries, 23

Calabi-Yau manifold, 183
canonical bundle, 107, 183
canonical bundle of $\mathbb{P}^n$, 188
canonical divisor, 36, 47
canonical divisor class, 47, 100
canonical homology bases, 28
canonical line bundle, 100
canonical sheaf, 100
Cauchy-Riemann differential equations,
    10
Cech cocycles, 95
Cech cohomology, 95
Cech cohomology groups, 97
chains, 22
Chern character, 127
Chern classes, 126
Chern classes of a manifold, 183
Chern number, 99
Chow ring, 125
Chow's theorem, 65
closed differential forms, 49
closed sets of $\mathbb{P}^1$, 62
coarse moduli space, 73
cocycle conditions, 88
coherent sheaves, 94
cohomologous cocycles, 89
cohomology of a complex, 175
cohomology sequence, 98
compact manifold, 8
compact Riemann surface, 13
compactification, 79
complete intersections, 191
complete prime ideals, 150

Theoretical and Mathematical Physics

An Introduction to Riemann Surfaces, Algebraic Curves and Moduli Spaces
2nd enlarged edition
By M. Schlichenmaier

Quantum Probability and Spectral Analysis of Graphs
By A. Hora and N. Obata

From Nucleons to Nucleus
Concepts of Microscopic Nuclear Theory
By J. Suhonen

Concepts and Results in Chaotic Dynamics: A Short Course
By P. Collet and J.-P. Eckmann

The Theory of Quark and Gluon Interactions
4th Edition
By F. J. Ynduráin

From Microphysics to Macrophysics
Methods and Applications of Statistical Physics
Volume I, Study Edition
By R. Balian

Titles published before 2006 in *Texts and Monographs in Physics*

The Statistical Mechanics of Financial Markets
3rd Edition
By J. Voit

Magnetic Monopoles
By Y. Shnir

Coherent Dynamics of Complex Quantum Systems
By V. M. Akulin

Geometric Optics on Phase Space
By K. B. Wolf

General Relativity
By N. Straumann

Quantum Entropy and Its Use
By M. Ohya and D. Petz

Statistical Methods in Quantum Optics 1
By H. J. Carmichael

Operator Algebras and Quantum Statistical Mechanics 1
By O. Bratteli and D. W. Robinson

Operator Algebras and Quantum Statistical Mechanics 2
By O. Bratteli and D. W. Robinson

Aspects of Ergodic, Qualitative and Statistical Theory of Motion
By G. Gallavotti, F. Bonetto and G. Gentile

The Frenkel-Kontorova Model
Concepts, Methods, and Applications
By O. M. Braun and Y. S. Kivshar

The Atomic Nucleus as a Relativistic System
By L. N. Savushkin and H. Toki

The Geometric Phase in Quantum Systems
Foundations, Mathematical Concepts, and Applications in Molecular and Condensed Matter Physics
By A. Bohm, A. Mostafazadeh, H. Koizumi, Q. Niu and J. Zwanziger

Relativistic Quantum Mechanics
2nd Edition
By H. M. Pilkuhn

Physics of Neutrinos
and Applications to Astrophysics
By M. Fukugita and T. Yanagida

High-Energy Particle Diffraction
By E. Barone and V. Predazzi

Foundations of Fluid Dynamics
By G. Gallavotti

Many-Body Problems and Quantum Field Theory An Introduction
2nd Edition
By Ph. A. Martin, F. Rothen, S. Goldfarb and S. Leach

Statistical Physics of Fluids
Basic Concepts and Applications
By V. I. Kalikmanov

Statistical Mechanics A Short Treatise
By G. Gallavotti